Oil Revenues and Accelerated Growth

Oil Revenues and Accelerated Growth

Absorptive Capacity In Iraq

Kadhim A. Al-Eyd

PRAEGER

PRAEGER SPECIAL STUDIES • PRAEGER SCIENTIFIC

Library of Congress Cataloging in Publication Data

Al-Eyd, Kadhim A
   Oil revenues and accelerated growth.

   Bibliography: p.
   1. Petroleum industry and trade--Iraq--Finance.
2. Iraq--Economic conditions. I. Title.
HD9576.I72A644   338.2'3   79-18596
ISBN 0-03-053306-6

Published in 1979 by Praeger Publishers
A Division of Holt, Rinehart and Winston/CBS, Inc.
383 Madison Avenue, New York, New York 10017 U.S.A.

© 1979 by Praeger Publishers

All rights reserved

1234567890 038 098765432

Printed in the United States of America

for

nadia and ali

with love

ACKNOWLEDGMENTS

This study is a revised and expanded version of a doctoral dissertation that was submitted to the Department of Economics at George Washington University, Washington, D.C. I am deeply indebted to the director of my research, Professor Ching-Yao Hsieh, and to Professors Mary Alida Holman and Charles Todd Stewart. I benefited from their constructive criticism and wise counsel.

I am very grateful to Dr. Ahmed Al-Samarrie of the U.S. Office of Management and Budget and Dr. Paul M. Meo of the World Bank staff for their thorough review of the manuscript and for the many challenging and stimulating discussions of some of the ideas contained in this study.

I owe a special word of gratitude to Professor Abbas Alnasrawi of the University of Vermont and Dr. Mehdi M. Ali of the World Bank staff for the perceptive comments they so generously offered.

Thanks are due to Professor Salih Nur Neftci of George Washington University for his help with the quantitative sections of this study and to Tamar Katz of the World Bank Staff for her assistance with the computer work involved.

Nabila Salah and Brandy Wilson typed the final manuscript with exceptional efficiency. I am very thankful to both of them.

My wife, Kathleen, skillfully and cheerfully typed several earlier drafts of this study. No words can express my appreciation of her unfailing support, encouragement, and happy sharing in this endeavor.

The responsibility for any errors and shortcomings in the book is solely mine. This study is the result of my own initiative, and it should in no way be construed as reflecting the views of the Iraqi government or the International Monetary Fund.

CONTENTS

	Page
ACKNOWLEDGMENTS	vii
LIST OF TABLES AND FIGURES	xi
LIST OF ACRONYMS	xiv

Chapter

1	INTRODUCTION	1
	Notes	8
2	THE ROLE OF THE OIL SECTOR IN THE IRAQI ECONOMY	10
	From Concession to Nationalization	10
	Concessions	10
	Difficulties	14
	Prelude to Nationalization	18
	Nationalization	20
	Production, Refining, and Exports	24
	Production	24
	Refining	27
	Exports	28
	The Impact of the Oil Sector	28
	Impact on the Balance of Payments	29
	Oil Revenues and Public Finance	31
	Oil Revenues and Economic Development	32
	Conclusion	35
	Notes	35
3	DEVELOPMENT PLANNING IN IRAQ: SURVEY AND EVALUATION	41
	Survey of Development Planning	41
	The Development Board	42
	Partial National Planning	44
	Comprehensive National Planning	46

Chapter		Page
	Evaluation of Development Planning	52
	The Strategy of Economic Development	52
	Mobilization and Allocation of Resources	55
	Plan Implementation	58
	Summary and Conclusion	58
	Notes	60
4	PERFORMANCE OF THE IRAQI ECONOMY UNDER DEVELOPMENT PLANNING	65
	Overall Performance	65
	Sectoral Performance	67
	Agriculture	67
	Industry	70
	Transport and Communications	75
	Summary	76
	Notes	77
5	ABSORPTIVE CAPACITY OF THE IRAQI ECONOMY	80
	The Concept of Absorptive Capacity	80
	Definition	80
	Sectoral Absorptive Capacity	82
	Measurement	83
	Limitations	86
	The Case of Iraq	89
	Gross Domestic Fixed Capital Formation	89
	Definition of Absorptive Capacity	92
	Measurement of Absorptive Capacity	93
	Notes	100
6	THE IMPACT OF THE RISE IN OIL PRICES ON THE ABSORPTIVE CAPACITY OF THE IRAQI ECONOMY	102
	Actual versus Extrapolated Domestic Investment	102
	The Terms of Trade Effect	106
	Other Probable Causes	115
	Foreign Labor	115
	Lower Rate of Return	116
	Intersectoral Impact	119
	Sovereignty Over Natural Resources	119
	Summary	120
	Notes	120

Chapter		Page
7	PROSPECTS FOR ACCELERATED ECONOMIC GROWTH IN IRAQ	123
	Absorptive Capacity and Accelerated Growth	123
	Theoretical Discussion	123
	Growth Rates Before and After the Rise in Oil Prices	126
	Necessary Conditions to Sustain Accelerated Growth	128
	Development Policies	128
	Oil Revenues	135
	Oil and Prospects for Accelerated Growth	136
	Conclusion	137
	Notes	138
8	OPEC: PERFORMANCE AND PROSPECTS	140
	The OPEC Experience	141
	Future Prospects	146
	Demand for OPEC Oil	147
	Oil Prices	149
	Notes	151
	Appendix A: NOTES ON DATA	153
	Appendix B: ORDINARY LEAST SQUARES METHOD	156
	Appendix C: TERMS OF TRADE INCOME EFFECT	158
	Appendix D: STATISTICAL TABLES	161
	BIBLIOGRAPHY	180
	ABOUT THE AUTHOR	190

LIST OF TABLES AND FIGURES

Table		Page
2.1	Iraq: Crude Petroleum Production and Export	25
2.2	Crude Petroleum Production in Selected Countries	26
2.3	Iraq: Sectoral Contribution to Gross Domestic Product	29
2.4	Iraq: The Structure of Employment	30
2.5	Iraq: Proportion of Oil Revenues to Total Government Revenues	33
2.6	Iraq: Revenues of Economic Development Programs and Plans, 1951-74	34
3.1	Iraq: Planned versus Actual Development Expenditures, 1951-58	44
3.2	Iraq: Planned versus Actual Development Expenditures, 1959-64	46
3.3	Iraq: The Five-Year Economic Plan, 1965-69	47
3.4	Iraq: Central Government Actual Allocations and Expenditures, 1965-69	48
3.5	Iraq: National Development Plan (1970-74) — Distribution of Total Investment	50
3.6	Iraq: National Development Plan (1970-74) — Central Government Annual Investment Programs	51
3.7	Iraq: Central Government Actual Allocations and Expenditures, 1970-74	52
3.8	Iraq: Sectoral Distribution of the Central Government Actual Investment Expenditures, 1951-74	53
3.9	Iraq: Indexes of Plan Implementation, 1951-74	59
4.1	Aggregate and Sectoral Real Growth Rates of the Iraqi Economy	66

Table		Page
5.1	Iraq: Gross Domestic Fixed Capital Formation at Current Prices, 1953-73	90
5.2	Iraq: Gross Domestic Fixed Capital Formation at Constant 1969 Prices, 1953-73	91
5.3	Aggregate and Sectoral Growth Rates of Gross Domestic Fixed Capital Formation in Iraq	96
6.1	Iraq: Gross Domestic Fixed Capital Formation, 1974-76	103
6.2	Iraq: Actual versus Extrapolated Gross Domestic Fixed Capital Formation at Constant 1969 Prices	104
6.3	Export Price Indexes in Iraqi Dinars	109
6.4	Iraq: Terms of Trade Income Effect	109
6.5	Iraq: Real Product and Real Income at Constant 1969 Prices, 1969-76	110
7.1	Aggregate and Sectoral Real Rates of Growth of the Iraqi Economy Before and After the Rise in Oil Prices	127
7.2	Iraq: Actual Expenditures of the National Development Plan (1973-76)	128
8.1	Changes in Real Output, 1974-78	142
8.2	Major Oil Exporting Countries: Changes in Real GDP	145
D.1	Iraq: Current Account, 1950-74	162
D.2	Iraq: Gross Domestic Product by Economic Sector at Current Prices, 1953-76	165
D.3	Iraq: Gross Domestic Product by Economic Sector at Constant 1969 Prices, 1953-76	168
D.4	Iraq: Gross Domestic Fixed Capital Formation by Economic Sector at Current Prices, 1953-75	171

Table		Page
D.5	Iraq: Gross Domestic Fixed Capital Formation by Economic Sector at Constant 1969 Prices, 1953-75	174
D.6	Results of Ordinary Least Squares Estimations of Aggregate and Sectoral Growth Rates of the Iraqi Economy	177
D.7	Results of Ordinary Least Squares Estimations of Aggregate and Sectoral Growth Rates of Gross Domestic Fixed Capital Formation in Iraq	178

Figure		
1	Absorptive Capacity	84
2	Impact of TTE on Aggregate Demand	113
3	Impact of TTE on Absorptive Capacity	114
4	Impact of a Lower Rate of Return on Absorptive Capacity	118

LIST OF ACRONYMS

API	American Petroleum Institute
BOD	British Oil Development Company
BPC	Basrah Petroleum Company
CBI	Central Bank of Iraq
CFP	Compagnie Française de Pétroles
CSO	Central Statistical Organization
DB	Development Board
DEP	Detailed Economic Plan
ECAFE	Economic Commission for Asia and the Far East
FAO	Food and Agricultural Organization
GDFC	gross domestic fixed capital formation
GDP	gross domestic product
GDR	German Democratic Republic
GDY	real income
GNP	gross national product
IBRD	International Bank for Reconstruction and Development
IMF	International Monetary Fund
INOC	Iraqi National Oil Company
IPC	Iraq Petroleum Company
MPC	Mosul Petroleum Company
NDP	National Development Plan
OPEC	Organization of Petroleum Exporting Countries
PEP	Provisional Economic Plan
TPC	Turkish Petroleum Company
TTE	terms of trade effect

Oil Revenues and Accelerated Growth

1

INTRODUCTION

According to W. W. Rostow the "take-off" stage is historically traceable to "a particular sharp stimulus."[1] The stimulus can take several forms including "a newly favorable international environment, such as the opening of British and French Markets to Swedish timber in the 1860s or a sharp relative rise in export prices."[2]

A new and substantially increased price for crude oil went into effect on January 1, 1974. That event was the culmination of a process that began in 1970 and reached its climax during the last quarter of 1973. The posted price of a barrel of Arabian Light 34° American Petroleum Institute (API) (the benchmark crude oil) rose from $1.80 on January 1, 1971 to $3.01 on October 1, 1973, to $5.12 on October 16, 1973, and to $11.65 on January 1, 1974. Given the magnitude of the increase in the price of crude oil, the rather low price elasticity of demand for crude oil in the short and medium terms, and the general expectation that oil prices will remain more or less constant in real terms over the next decade, the event logically can be regarded as a sharp stimulus, which ought to trigger the take-off stage in the oil exporting countries.[3]

Such an outcome is by no means automatic. Sustained economic growth requires among other things the conversion of rising domestic savings into domestic investment in the various sectors. Historical experience shows that the development of major export industries is not necessarily followed by self-sustaining growth. Rostow pointed out that, "Enlarged foreign-exchange proceeds have been used in many familiar cases to finance hoards (as in the famous case of Indian bullion imports) or unproductive consumption outlays."[4] In addition to hoarding and unproductive outlays, Rostow could have cited other obstacles such as the limits that restrict the absorptive

1

capacity of the economy. He did, however, acknowledge this fact, albeit indirectly, when he spoke of the social change that is essential for a successful take-off.[5]

Most oil exporting countries are well poised to undertake the task. A key element in their overall strategy must be the transformation of the oil sector from an "enclave" into a "leading" sector.[6] There are theoretical as well as empirical reasons that argue in favor of an early transformation of this vital sector. On the empirical level, it is important to note that in most cases the rise of a raw material export industry "was a stage at which the country remained for decades or even centuries, while the characteristic pattern of export economy development blocked further advance to the stage of a balanced economy."[7]

Oil exporting countries are no exception. A case in point is Venezuela. On August 25, 1954, the Office of Intelligence Research of the Department of State, Washington, D.C., released a report containing estimated ratios of gross investment to gross national product (GNP) for several countries including Venezuela.[8] Although the estimated ratio for Venezuela was given as 23 percent of the GNP, its economy was judged to be still attempting a take-off.[9] Rostow's explanation of this anomalous situation was that "Venezuela has been for some time an 'enclave economy,' with high investment-rate concentrated in a modern export sector whose growth did not generate general economic momentum in the Venezuelan economy."[10]

The fact that enclave sectors have existed for a long time in the midst of traditional economies is attributable, among other things, to their failure to contribute to the expansion of local markets. Enclave sectors not only have few linkages with the local economy, but they also have few spread affects. Owing to their employment and wage policies, they fail to raise incomes significantly enough to have an impact on aggregate demand and spread the money economy. This characteristic of enclave sectors was stated succinctly by Hla Myint:

> In recent times, it has become fashionable to look upon the mines and plantations as "foreign enclaves" which are incapable of transmitting a satisfactory type of economic growth to the developing countries. The critics tend to put most of the blame on the fact that they are concerned with primary production and not with manufacturing industry. Our analysis has, however, suggested that the failure of the mines and plantations to become the "leading sector" in the underdeveloped countries was due, not to their producing primary exports as such, but to their cheap labour policy which

has perpetuated the pattern of low wages and low productivity. With the system of using indigenous labour mainly as an undifferentiated mass of brawn-power, it is not surprising that Adam Smith's vision of the growth of the exchange economy and the division of labour "improving the skill and dexterity of the people" should remain largely unfulfilled.[11]

Myint's conclusion is certainly valid as a general proposition. However, in the case of oil extraction, a capital intensive activity, it is not so much a question of employment and wages as much as it is a question of further industrialization of the oil sector. Refineries, petrochemical industries, and so on are important not only in terms of the employment they generate and the technology they bring but also in terms of their impact on the rest of the economy through linkages.

On the theoretical plane, it is helpful to recognize that there are two theories of take-off: aggregative and sectoral.[12] The aggregate theory is basically a Harrod-Domar growth model, where the rate of growth in national income exceeds the rate of population growth—hence, the emphasis on a sharp rise in the savings (investment) rate during the take-off stage.

The other theory is "a sectoral, non-linear, threshold notion. This is the realm of the leading sector, with its forward, backward, and spreading effects breathing regular innovation into the heretofore slumbering corpus economicum."[13] This notion derives from the theory of unbalanced growth. Albert O. Hirschman, for example, argues that in a situation where the limiting factor is the inducement or ability to invest, rather than lack of savings, the sectoral approach is more promising:[14]

> The ability to invest is acquired and increased primarily by practice; and the amount of practice depends in fact on the size of the modern sector of the economy. In other words, an economy secretes abilities, skills, and attitudes needed for further development roughly in proportion to the size of the sector where these abilities are already required and where these attitudes are being inculcated.[15]

It therefore follows that the development of the most modern sector—that is, the oil sector—should assist the oil exporting countries in two ways: (1) expansion of their domestic markets through linkages, rise in incomes, and concomitant increase in aggregate demand and (2) acceleration of investment in other sectors. The latter possibility will be given momentum by the spillover of "abilities, skills, and attitudes" from the oil sector to the rest of the economy.

This observation takes us straight back to our original observation—namely, in order for the oil exporting countries to benefit from the sharp stimulus and enter an era of sustained growth, they must launch a successful attack on the limits that restrict their absorptive capacity.

The conventional wisdom is that the absorptive capacity of the oil exporting countries is limited. For example, in 1974 the International Bank for Reconstruction and Development (IBRD) estimated that the cumulative reserves of the members of the Organization of Petroleum Exporting Countries (OPEC) would amount to some $650 billion in 1980 and $1.2 trillion by 1985.[16]

In terms of investment theory, the projected surplus means that in the short term domestic savings will exceed domestic investment in the oil exporting countries. At first the domestic resource surplus manifests itself as a current account surplus. Initially, actual developments were in conformity with this expectation. The combined current account surplus of the major oil exporting countries amounted to $6 billion in 1973, $67 billion in 1974, $35 billion in 1975, and $41 billion in 1976.[17] After peaking in 1974, the current account surplus of OPEC has exhibited a downward trend. Some have attributed this development to a sudden improvement in the ability of these countries to invest:

> The speed with which highly ambitious development strategies could be formulated and implemented in the oil exporting countries during the period after 1973 was not anticipated by many commentators in the early days of the oil crisis. Largely as a result of these factors, the balance of payments surplus on current account of the major oil exporting countries declined from $68 billion in 1974 to $35 billion in 1977; and it has been projected to decline further, to around $10 billion in 1978.[18]

Although the rapid decline in the current account surplus of the oil exporting countries could be indicative of expansion in their absorptive capacity, it is not prudent to conclude so only on the strength of this evidence. If anything, history teaches that the presence of a class of "luxury importers" has always been one of the major obstacles that impedes the balanced growth of export economies.[19] The interest here is in the effective—that is, efficient utilization—of oil revenues. Unproductive consumption such as luxury imports and military purchases do not signify improvement in the "ability to invest" nor do they portend accelerated growth. These considerations provided the motivation for this study.

Oil exporting countries differ with respect to the size of their population, their other natural endowments, and the degree of their commitment to the cause of economic development. That is to say, what is true for any one of them is not necessarily true for the rest of them. It is for this reason that members of OPEC have been classified into low and high absorbers.* Therefore, it is both logical and practical to concentrate on a single oil exporting country. The purpose of this study is to investigate the impact of the rise in oil prices on the absorptive capacity of the Iraqi economy, examine the utilization of oil revenues in general and within the context of the country's development policy in particular, and assess future prospects for accelerated growth in Iraq. The question under investigation here is: Did the sharp rise in the price of crude oil have a positive impact on the absorptive capacity of the Iraqi economy?

Absorptive capacity has been defined as "that amount of investment or that rate of gross domestic investment expressed as a proportion of GNP, that can be made at an acceptable rate of return, with the supply of co-operant factors considered as given."[20] Evidently, absorptive capacity is governed by two main factors: acceptable rate of return and supply of cooperant factors. The criterion of rate of return, clearly, owes its origin to Keynes's "marginal efficiency of capital," a fact that reduces the concept of absorptive capacity to a schedule relating investment to the rate of return. In other words, the lower the rate of return acceptable to the investor—the body responsible for making investment decisions—the higher the level of investment and, hence, the larger the absorptive capacity of the economy.

The definition contains no reference to any particular rate of return. The omission is deliberate. For instance, limiting its meaning to financial rate of return would imply that the cutoff rate is the average long-term rate of investment prevailing outside the economy. Put differently, the absorptive capacity of the economy is that amount of invested capital that sets the rate of return exactly at the rate available outside the economy. This is a restrictive interpretation. It is preferable to think in terms of a social rate of return in order to account for complementarity and indivisibility of

---

*The low absorbers are: Abu Dhabi, Kuwait, Libya, Qatar, and Saudi Arabia; the high absorbers are: Algeria, Indonesia, Iran, Iraq, Nigeria, and Venezuela.

investment projects and the external economies and diseconomies associated with their undertaking.[21]

The constancy of the supply of cooperant factors (skilled labor, managerial and entrepreneurial talents, and so on) assumption implies that investment activity could come to a complete halt upon the exhaustion of any cooperant factor. This assumption denies the possibility of factor mobility or substitution and is only valid within the context of a closed economy. Without explaining why, John H. Adler states that increasing the supply of cooperant factors in the short run "is either a physical impossibility or is so costly that it reduces the return on capital below the acceptable rate."[22]

Again, this is a rigid assumption. Absorptive capacity can be expanded in the short run by allowing investment to proceed at higher costs by deliberately accepting higher capital-output ratios in cases where capital can be substituted for skilled labor when the latter is in short supply or by importing foreign labor. Thus, recognizing the principle of substitution and the present international setting where the mobility of factors of production is more or less free, it would be difficult to justify the proposition that the cost of improving the economy's absorptive capacity would be prohibitive.

To summarize the discussion up to this point, the situation depicted so far is one where: (1) potential domestic savings are likely to exceed domestic investment requirements; (2) the planner, though rational, is not a profit maximizer in the sense one views entrepreneurs to be and is, therefore, likely to choose a low financial rate of return; and (3) cooperant factors can be augmented through importation or substitution by other factors in relative abundance. These considerations lead to the hypothesis that the quadrupling of oil prices and the concomitant increase in oil revenues should have a positive impact on the absorptive capacity of the Iraqi economy. This study endeavors to test this hypothesis—that is, to ascertain whether there has been expansion in the absorptive capacity of the Iraqi economy following the rise in oil prices and by how much. The study also aims to explain the factors responsible for such developments and assess both the prospects for sustaining the new level of investment and its implication for the future growth of the economy.

Testing this hypothesis involves two steps. First, it is essential to adopt a measure of absorptive capacity. H. B. Chenery's and A. MacEwan's observations led them to conclude that the level of investment in any given period could not exceed the level of investment in the preceding period that was multiplied by an exogenously determined growth rate in the range of 15 to 20 percent.[23] According to this formulation, the absorptive capacity limit takes the form of a maximum ceiling on the level of investment in any given period.

It is proposed that the absorptive capacity of the Iraqi economy be measured in a similar fashion. It is postulated that the observed rate of growth of domestic investment in Iraq during the period preceding the rise in oil prices is a maximum rate determined by the limit of absorptive capacity of the economy. This postulate is premised on the empirical observation that domestic savings consistently had exceeded domestic investment throughout the period from 1950 to 1973. Since Iraq did not suffer from shortages in investable funds, its failure to achieve higher levels of domestic investment must logically be attributed to the limits of its absorptive capacity. This conclusion is corroborated by the fact that a reasonable effort to expand the country's absorptive capacity was exerted as indicated by the consistent surplus of actual revenues earmarked for public investment over actual public investment expenditures. Other evidence includes the absence of any serious effort on the part of the planner to mobilize private domestic savings as well as foreign savings (public or private) for domestic investment purposes. This lack of enthusiasm is understandable given the inability to exhaust public savings at the disposal of the authorities.

Second, in order to demonstrate that the rise in the price of crude oil had a positive impact on the absorptive capacity of Iraq, it must be shown that an upward shift in investment had in fact occurred. It is proposed, therefore, to compare the actual level of investment during the years following the rise in oil prices with the expected (that is, extrapolated) level of investment for those years. In order to evaluate the implications of the change in the level of investment, if any, it is proposed to repeat the test for public and private investment and for investment in agriculture and industry.

The analysis will be based on data gathered from Iraqi sources, particularly the Ministries of Planning and Finance, the Central Statistical Organization (CSO), and the Central Bank of Iraq (CBI), and from international organizations, particularly the International Monetary Fund (IMF) and the IBRD. Various tools of economic analysis, including both descriptive and inferential statistics, will be used as necessary.

Throughout this book public documents are referred to by title and number. Iraqi documents were published in Baghdad, Iraq, by the agency listed in the reference, one year after the year indicated on title. IMF documents were published in Washington, D.C., by the International Monetary Fund, the same year as indicated on title. U.N. documents were published in New York, by the U.N., two years after the year indicated on the title.

NOTES

1. W. W. Rostow, The Stages of Economic Growth (New York: Cambridge University Press, 1960), p. 36.
2. Ibid., p. 37.
3. As is well known, W. W. Rostow's generalizations about the process of economic growth are controversial. Quotes from his famous work do not necessarily imply an endorsement of his theory but simply indicate that the take-off stage he theorized is relevant to the theme of this book and, consequently, to the discussions that will follow. For criticism of Rostow's theory the reader is referred to papers presented at the International Economic Conference and published in W. W. Rostow, ed., Economics of Take-Off into Sustained Growth (New York: St. Martin's Press, 1963); Kenneth Berrill, "Historical Experience: The Problem of Economic 'Take-Off,'" in Economic Development with Special Reference to East Asia, ed. Kenneth Berrill (London: Macmillan, 1965), chap. 7; P. A. Baran and E. J. Hobsbawm, "The Stages of Economic Growth," Kyklos 14 (1961): 234-43; P. T. Bauer and Charles Wilson, "The Stages of Growth," Economica 29 (1962): 190-200; S. G. Checkland, "Theories of Economic and Social Evolution: The Rostow Challenge," Scottish Journal of Political Economy 7 (1960): 169-93; D. C. North, "A Note on Professor Rostow's 'Take-Off' into Self-Sustained Economic Growth," The Manchester School of Economic and Social Studies 26 (1958): 68-75; Goran Ohlin, "Reflections on the Rostow Doctrine," Economic Development and Cultural Change 11 (1961): 648-55; and Henry Rosovsky, "The Take-Off into Sustained Controversy," The Journal of Economic History 25 (1965): 271-75.
4. Rostow, Stages of Economic Growth, p. 49.
5. Ibid., chap. 3.
6. On enclave sectors see Hla Myint, The Economics of the Developing Countries (New York: Praeger, 1965), chap. 4. On the lack of benefits of foreign investments in export industries see Hans Singer, "The Distribution of Gains between Investing and Borrowing Countries," American Economic Review 40 (1950): 473-85; and Jonathan V. Levine, The Export Economies: Their Pattern of Development in Historical Perspective (Cambridge, Mass.: Harvard University Press, 1960). For a contrasting view see Raymond Vernon, "Foreign-Owned Enterprise in the Developing Countries," in Economics of Trade and Development, ed. James D. Theberge (New York: John Wiley and Sons, 1968), pp. 446-63. On leading sectors see Albert O. Hirschman, The Strategy of Economic Development (New Haven, Conn.: Yale University Press, 1958), chap. 6; and W. W. Rostow, "Leading Sectors and the Take-Off," in The Economics of Take-Off into Sustained Growth, ed. W. W. Rostow (New

York: St. Martin's Press, 1963), pp. 1-21. For more recent contributions see Lauchlin Currie, "The 'Leading Sector' Model of Growth in Developing Countries," Journal of Economic Studies 1 (1974): 1-16; and James A. Hanson, "The Leading Sector Development Strategy and the Importance of Institutional Reform: A Reinterpretation," Journal of Economic Studies 3 (1976): 1-12.

7. Levin, Export Economies, pp. 2-3.
8. Rostow, Stages of Economic Growth, p. 44, fn.
9. Ibid.
10. Ibid.
11. Myint, Economies of the Developing Countries, p. 64.
12. Albert Fishlow, "Empty Economic Stages?" Economic Journal 75 (1965): 112-25.
13. Ibid.
14. Hirschman, Strategy of Economic Development.
15. Ibid., p. 36.
16. Rimmer de Vries, "The Build-Up of OPEC Funds," World Financial Markets (New York: Morgan Guaranty Trust Company, September 23, 1974), p. 1.
17. International Monetary Fund, Annual Report 1977, p. 13.
18. David R. Morgan, "Fiscal Policy in Oil Exporting Countries, 1972-78," IMF Staff Papers 26 (1979): 55-56.
19. Levin, Export Economies, pp. 177-78.
20. John H. Adler, Absorptive Capacity, The Concept and Its Determinants (Washington, D.C.: Brookings Institution, 1965), p. 5.
21. Ibid., pp. 9-15.
22. Ibid., p. 5.
23. H. B. Chenery and A. MacEwan, "Optimal Patterns of Trade and Aid: The Case of Pakistan," in The Theory and Design of Economic Development, ed. Irma Adelman and Erik Thorbecke (Baltimore: Johns Hopkins University Press, 1966), pp. 151-52.

2

THE ROLE OF
THE OIL SECTOR IN
THE IRAQI ECONOMY

Iraq is an Arab country in western Asia with an area of about 171,000 square miles and a population of a little over 12 million.[1] Cultivable land is estimated at 48 million <u>meshara</u> (approximately 30 million acres).[2] The ratio of cultivable land per person is comparatively large and is thought to be in excess of what the country could support (provided the land is developed) or what it actually supported in the past during the golden ages of Mesopotamian civilizations.[3]

Next to land, oil is the most important natural resource. Iraq is a major oil producing country and a founding member of OPEC. In 1978 Iraq's crude oil production accounted for approximately 12.4 percent of total Middle Eastern production and 4.4 percent of total world production.[4]

FROM CONCESSION TO NATIONALIZATION

Concessions

Knowledge of the existence of oil in Iraq dates back to ancient times. Indeed, surface indications of the presence of oil in all of western Asia have been there throughout human history.[5] However, it was not until the latter part of the nineteenth century that Western interests became aware that Iraq contained rich oil fields and began their efforts to secure concessions from the Ottoman authority for the commercial exploitation of Iraqi oil.[6] The possibility that rich mineral deposits existed prompted Sultan Abdul Hamid to confer, "by special firmans, the concessions of the vilayets of Mosul and

Baghdad, respectively, on his Civil List, thus making the concessions his private business."[7]

In 1904 the Anatolian Railway Company, a German concern acting for the Deutsche Bank, and the Civil List concluded an agreement permitting the former to survey the vilayets (provinces) of Mosul and Baghdad, with an option to enter into contract with the Civil List for joint operation of discovered oil fields.[8] Two years later the Civil List terminated its agreement with the Anatolian Railway Company and "began negotiations with William Knox D'Arcy, who had the active support of the British Ambassador at Constantinople.[9]

By 1912 the number of interested parties in Iraqi oil rose to four, namely, the Deutsche Bank (German), the D'Arcy group (British), the Anglo-Saxon Company, a subsidiary of the Royal Dutch-Shell group (British-Dutch), and the Colby M. Chester group (American).[10] Also in that year, political and practical considerations converged and resulted in the formation of the Turkish Petroleum Company (TPC). First, the Europeans were united in their opposition to U.S. entry; second, the British and Germans were very close to reaching an agreement on the Baghdad railway project; and third, the British did not oppose the efforts of the Anglo-Saxon Company because the Dutch group was to a considerable extent under British control.[11]

Against this background the TPC was established as a British joint stock company for the purpose of acquiring all claims to the Iraqi oil fields. Its shares were distributed as follows: the Turkish National Bank, a British-owned Bank, 50 percent; the Deutsche Bank, 25 percent; and the Anglo-Saxon Company, 25 percent.[12] On March 19, 1914, representatives of the British government, the German government, the Turkish National Bank, the Royal Dutch-Shell Company, and the D'Arcy group signed what is popularly known as the Foreign Office Agreement (so called because it was signed in the British Foreign Office).[13]

Under the terms of the Foreign Office Agreement the capital of the TPC was increased from £80,000 to £160,000; the D'Arcy group, represented by the Anglo-Persian Oil Company, took over the interest of the Turkish National Bank, whereas the shares of the Deutsche Bank and the Anglo-Saxon Company remained unchanged at 25 percent each.[14] An Armenian by the name of Sarkis Colouste Gulbenkian was rewarded with 5 percent beneficiary interests (2.5 percent from D'Arcy and 2.5 percent from Anglo-Saxon) for his help in reconciling the British and German interests, and more important, for his unceasing efforts on behalf of the TPC to secure a concession from Turkish authority.[15]

Five days after the signing of the Foreign Office Agreement, "the Ambassadors of Germany and Great Britain, in identical notes to the Grand Vizier, requested him to grant the Turkish Petroleum Company a concession for the exploitation of the oil fields in the vilayets of Mosul and Baghdad."[16] The grand vizier (prime minister) replied to the German ambassador in Constantinople on June 28, 1914, offering a vague promise to grant the requested concessions.[17] Fulfillment of this promise was delayed, however, due to the outbreak of World War I.

During the war years the French and British governments maintained an ongoing dialogue concerning Iraqi oil; finally, on April 24, 1920, they signed the San Remo Oil Agreement according to which the German interest in TPC was transferred to France.[18]

In 1923 the TPC approached the Iraqi government, in its capacity as successor to Ottoman authority in the area, demanding the concession promised to it by the grand vizier.[19] The newly organized government denied that the letter of the grand vizier to the German ambassador constituted a valid concession and issued a statement authorizing the minister of finance to negotiate new concessional terms with TPC.[20] It should be pointed out that in the opinion of many, including that of the U.S. Department of State, the grand vizier's letter was not legally valid.[21] Writing almost 40 years later, Benjamin Shwadran stated, "As to the concession of the Turkish Petroleum Company, it cannot be claimed that the Grand Vizier's letter of June 28, 1914, was a legally valid concession which could have been protected by the peace treaty with Turkey."[22]

After prolonged negotiations the government of Iraq, then under British mandate, signed the TPC concession on March 14, 1925. The signature of the government was obtained under duress. The government's request for a modification in the terms of the concession, drafted by the British high commissioner in Baghdad, in order to allow for local participation, was rejected by the TPC.[23] Instead, pressure was exerted on the newly established state through the utilization of the problems of the disposition of the Mosul vilayet, which according to the Sykes-Picot Agreement of May 16, 1916, was to be included in the projected French sphere of influence.[24]

To avert the loss of Mosul, the government acceded to the demand of the TPC and signed the concession agreement. The ministers of justice and education resigned in protest because the 20 percent state participation stipulated in the San Remo Oil Agreement of April 24, 1920, was not honored in this concession.[25] The British authorities deliberately delayed the ratification of the Iraqi constitution until after the signing of the concession.[26] The reason for this was that, after the ratification of the constitution, the Council of

Ministers would have no power to grant concessions in connection with any of the country's natural resources except by special law. This, the British authority wanted to avoid, since passage of such a law would have met with determined resistance.[27]

The concession agreement gave TPC an exclusive right to explore, produce, and market oil in all of Iraq except the vilayet of Basrah and the "transferred territories" for a period of 75 years. In return the company agreed to pay the government a royalty of Sh 4 (gold) per ton of crude oil produced.[28]

U.S. interest in Iraqi oil dated back to 1908 when Rear Admiral Colby M. Chester entered into negotiations with the Turkish government for the purpose of obtaining railway and mining concessions.[29] On March 9, 1910, the Turkish minister of public works signed a concession granting the Ottoman-American Development Company, which was organized by Chester, among other things, the right to construct a railway line, "from some point on line 'a' to Suleimaniah and the Persian border via Mosul and Kirkuk. The concession granted mineral rights, including oil, for twenty kilometers on both sides of the line."[30] The concession was never ratified by the Turkish Parliament due to the outbreak of the Balkan wars and later World War I.[31]

When the news of the San Remo Agreement between France and Great Britain became known, the U.S. Department of State strongly protested the exclusion of U.S. companies and demanded an "open door" policy of oil development in Iraq and other parts of the now defunct Ottoman Empire.[32] Throughout most of the 1920s, negotiations and consultations were conducted simultaneously between U.S. and British oil companies, on the one hand, and between the U.S. Department of State and the British Foreign Office, on the other. Finally, on July 31, 1928, the Red Line Agreement (so called because the map attached to it was marked out by a red line) was signed:[33]

> The Red Line Agreement, variously assessed as a sad case of wrongful cartelization or as an enlightened example of international co-operation and fair sharing, was to hold the field for twenty years and in large measure determined the pattern and tempo of oil development over a great part of the Middle East.[34]

Under the terms of the Red Line Agreement, the Near East Development Corporation, a U.S. concern, acquired 23.75 percent interest in the TPC from the Anglo-Persian Oil Company, thus reducing the latter's shareholding to 23.75 percent.[35] The Near East Development Corporation was organized in February 1928 by five

U.S. companies. Three participants—namely, Gulf, Atlantic, and Mexican—subsequently sold their interest to Standard Oil of New Jersey and Standard Oil of New York; the latter became Socony-Vacuum Oil in 1931.[36] Today these two companies are known as Exxon and Mobile, respectively. On June 8, 1929, the name of the Turkish Petroleum Company was changed to Iraq Petroleum Company (IPC).[37] IPC limited its functions to production and transport and acted "in the capacity of non-profit-making supplier of cheap crude oil to its constituents, one of whom would buy the Gulbenkian share at a just valuation."[38]

The terms of the 1925 concession were modified on March 24, 1931, when a new agreement giving IPC exclusive right to exploit all lands east of the Tigris River, covering an area of 32,000 square miles, was signed.[39]

Another British concern (with Italian, French-Swiss, and German capital participation), the British Oil Development Company (BOD), succeeded in securing a concession from the Iraqi government on May 25, 1932.[40] The 75-year concession gave BOD the sole right to exploit a 46,000 square mile area in the vilayets of Mosul and Baghdad, all of which were situated to the west of the Tigris River and to the north of the thirty-third parallel. About nine years later (that is, in 1941) the parent companies of IPC completed the acquisition of BOD, formed the Mosul Petroleum Company (MPC), and transferred all the shares of the former to the latter.[41]

The part of Iraq not covered by previous concessions was leased on December 4, 1938, to the Basrah Petroleum Company (BPC), an affiliate of IPC.[42] The concession involved an area of 93,000 square miles for a period of 75 years. With this concession the IPC group sealed its control of all of Iraq, including the Iraqi interest in the Iraqi-Saudi Neutral Zone.[43]

Difficulties

In August 1950 IPC and its two affiliates, MPC and BPC, agreed to raise the royalty rate by 50 percent (that is, to Sh 6 per metric ton).[44] This, however, was only a temporary solution, which did not address the basic disagreement between the government and the IPC group on the valuation of the shilling. It will be recalled that the concessions provided for a payment of tonnage royalty of Sh 4 *gold* per metric ton. The companies interpreted this to mean that they were to pay the sterling equivalent of the price of gold as it was officially determined by the Bank of England. The companies' interpretation was never recognized by the government:

ROLE OF THE OIL SECTOR / 15

The Government had long requested clarification of the royalty-revision formulae provided in the concessions and was unlikely to await patiently the end of the twenty-year period for which a royalty of "4s. gold" per ton was prescribed. Nor had the "gold question" itself ever been settled; and the sums at stake in the controversy were all the greater since the war-time price of gold in Middle Eastern bazaars had far exceeded—had doubled or trebled—the "Bank of England price," upon which the companies based their calculations: if gold were to mean the sterling equivalent of gold sovereigns, according to the Baghdad money-changers' tariff, the Company's outstanding debt was formidable.[45]

There were other issues awaiting solution. First, there was the question of participation. As pointed out earlier, the San Remo Agreement of 1920 gave Iraqi interests the right to participate up to a maximum of 20 percent of the share capital of any company eventually authorized to exploit Iraqi oil.

Accordingly, the Iraqi negotiators "persistently demanded participation for the Iraqi Government in the share capital of the Company [TPC] in accordance with the provisions of the San Remo Agreement." For a long period, they "held out for a free block of shares in addition to royalty," and asked for the right to nominate an Iraqi director on the Company's board. . . . all in vain as it turned out.[46]

The TPC rejected the idea of local participation and the most Iraqi negotiators could secure was the following provision in the March 14, 1925 concession:

Whenever an issue of shares is offered by the company to the general public, subscription lists shall be opened in Iraq simultaneously with lists opened elsewhere, and Iraqis in Iraq shall be given preference to the extent of at least twenty per cent of such issue.[47]

TPC, or more accurately its shareholders, had not the slightest intention of complying with this provision. They took, or more accurately, created, the first opportunity to circumvent this provision without consulting the Iraqi government.[48]

The parent companies of IPC [successor of TPC] have established the latter as a "private company" under

British law. A private company is precluded by law from issuing any shares to the public. This means that Iraqi capital cannot participate, as all increases in capital are made directly by the parent companies.[49]

In addition to the gold and participation questions, there remained the outstanding issues "of increasing the output of Iraq oil to a higher minimum rate and the more rapid substitution of Iraqis in the higher positions in the industry. Negotiations on these matters, which were pressed by Iraqis for both economic and political reasons, occupied much of the attention of both parties between 1947 and 1952."[50]

While the Iraqi government and the companies were locked in these disputes, certain developments in the region evidently paved the way for a major alteration in the concessions of IPC and its two affiliates. On December 31, 1950, the Saudi government concluded an agreement with Aramco providing for a 50/50 profit-sharing arrangement; on April 30, 1951, the Iranian Majlis (Parliament) passed an Enabling Law providing for nationalization of oil, and the 50/50 profit-sharing agreement between the government of Kuwait and the Kuwait Oil Company went into effect on December 1, 1951.[51]

The Iraqi bargaining position was enhanced by these events as well as by the hostility toward the foreign oil companies exhibited by the press and public opinion. For example, in complete defiance of the government's wishes, the opposition members in the Iraqi Parliament introduced, on March 25, 1951, a resolution demanding the nationalization of the oil companies operating in Iraq. The resolution cited the companies' failure to increase production rates to match those of neighboring countries, their refusal to pay royalty in gold, and their failure to train Iraqi nationals on certain technical jobs, as sufficient grounds for nationalization.[52]

In this atmosphere of change and uncertainty the companies "agreed at once to negotiations on the basis of one of many possible 'half-and-half' formulas."[53] The government and the companies signed on February 3, 1952, and Parliament ratified late in that month an agreement providing for equal sharing of profits on crude oil production, calculated at seaboard value.[54] Although the agreement failed to satisfy all of the Iraqi demands,

> the Government was now assured that the production of Iraqi oil would, subject to <u>force majeure</u>, be maintained at the high level already made possible by new pipeline construction and oilfield development; the M.P.C. and I.P.C. together were to produce 22 million t/y from 1954 onwards and the B.P.C. to raise

its production from 2 million to 8 million t/y as from the end of 1955.[55]

Under this agreement, the government's oil revenues were projected to amount to not less than £ 30 million in 1953 and 1954 and not less than £ 50 million in 1955 and thereafter. Furthermore, the government was given the option of receiving part of its share in kind, up to 12.5 percent of total oil production for export or for resale to the company.[56]

Implementation of the 1952 agreement was accompanied by the emergence of new difficulties, thus lengthening the list of unresolved issues. A problem, which should have been anticipated and which proved later on to be a very thorny one indeed, was that of cost determination. Since the government's revenues depended on net profits, it wanted very much to ascertain that costs claimed by the companies were proper and actually incurred. In this connection, the government thought that the "fixed costs" claimed by the companies were too high; that the amortization of "dead rents," paid by the companies prior to the beginning of commercial production of oil, was not permissible, since they were not regarded by the government as recoverable expenditures; and that "grants" made by the oil companies were not deductible expenses, particularly since the government was not consulted regarding these grants.[57] Other cost-related complaints by the government included the treatment of a portion of the London office expenses as production costs, the accounting treatment of expenditures on public relations, and the accounting treatment of drilling and exploration expenditures that the government believed should be amortized over a period of 20 years.[58]

The other major problem concerned posted prices. Payments to the government, according to the 1952 agreement, were to be based on posted prices. On February 9, 1956, the parent companies of the IPC group announced a price cut of $0.05 per barrel of Iraqi crude exported from Fao in southern Iraq. The government protested this unilateral action, particularly since no similar reduction was announced for other Arabian Gulf crudes.[59] Another price cut was indirectly effected when, on September 23, 1957, the company introduced the stripping method, due, it was claimed, to world demand then for low gravity oil. The stripping method is

> a form of refining leading to an appreciable drop in the gravity of crude oil. . . . Posted prices . . . are a function of gravity: the higher the gravity of oil, the higher its price (2 cents per one degree API gravity). The Iraqi Government did not object to the stripping

18 / OIL REVENUES & ACCELERATED GROWTH

method as such, but wanted to be paid revenues based on posted prices as if oil exports of crude were made without any treatment.[60]

Price discounts were another source of dispute. It seems that the companies, under the pretext of encouraging exports, succeeded in concluding an agreement with the government on March 24, 1955, but retroactive to January 1, 1954, which provided for sales as well as incentive discounts.[61] Under the terms of this agreement, the company was granted

> a 2 per cent sales discount on exports, reportedly to cover owners' selling expenses in disposing of Iraqi oil. The incentive discounts on higher volumes of exports were 5 per cent on the first 8 million tons above the minimum of 30 million tons, 7.5 per cent on the next 8 million tons, and 10 per cent on exports beyond.[62]

In 1957 the government persuaded the companies to eliminate the incentive discounts "and to reduce sales discount from 2 to 1 per cent."[63]

Finally, there was the problem of natural gas. Operating companies in neighboring Iran granted the government's request that it be entitled to use domestically the surplus gas produced by these companies.[64] "The Iraqi Government also asked the companies, early in 1957, to renounce immediately their claim on natural gas in surplus to their own requirements."[65] It was not until 1961 that the companies finally agreed to provide the government with a specified quantity of natural gas and proposed the formation of a joint committee to "explore possibilities of utilizing any future amounts of gas that may become available over and above the amount guaranteed to the government."[66]

Prelude to Nationalization

A new government assumed power in Iraq after the July 14, 1958 revolution. The new government invited the IPC group to enter negotiations for the purpose of settling all outstanding issues—including relinquishment by the companies of the areas covered by the concessions but not actually exploited. This problem evidently was accorded special importance by the new government. "Retention by the companies of vast areas, all of oil potential, and leaving them either unexplored or utilized, involved the abuse of a right enjoyed by the companies which had to be eliminated."[67]

Other items on the government's agenda were familiar: calculation of production cost, method of fixing prices, cancellation of the price discount, surrender by the companies of surplus natural gas, increase of Iraq's share in oil profits, and Iraqi participation in the companies' capital.

The negotiations were protracted and fruitless and came to a standstill in October 1961. On December 13, 1961, Law No. 80 withdrew the companies' concession rights on 99.5 percent of the concession area and imposed certain cargo dues and port charges on oil shipments through Basrah Port.[68] These areas were later awarded to the Iraqi National Oil Company (INOC), which was established on February 8, 1964.[69]

The companies never accepted Law No. 80 and for the next ten years pursued a policy designed to abrogate it or, at least, go around it. Their immediate response was to slow down production in order to penalize the country by denying it much needed oil revenues. Production policies practiced by the oil companies caused the annual rate of growth in Iraqi crude oil production to drop from 22 percent during the 1950-60 period to a mere 4.9 percent during the 1960-70 period.

Reduced production did not only deprive the country from additional revenues but pushed the production cost per barrel up, thus reducing government take per barrel. The Ministry of Oil and Minerals estimated the financial losses to Iraq resulting from these production measures for the years from 1962 to 1970 to exceed ID 550 million.[70]

The companies' prejudicial policy also was manifested through their investment policy. After reaching a peak of ID 22.8 million in 1960, the companies' investment expenditures dropped to ID 4.7 million in 1962 and ID 1.3 million in 1963. The average annual gross investment between 1964 and 1969 was ID 565,000. The companies felt no embarrassment despite the fact that during the same period their average annual share in the profits realized from crude oil extraction exceeded ID 140.7 million, all of which was repatriated. Comparison between gross investment expenditures and profits reveals that the ratio of the former to the latter was a minute 0.4 percent annually.[71]

The discrimination against Iraq was pursued with vigor. In 1964 the oil companiés reached an agreement with OPEC providing for royalty expensing. The IPC group offered to expense royalty to Iraq, but the offer was rejected by the government because it was bound by unreasonable conditions, the most objectionable of which was the insistence by the companies on compulsory arbitration for the settlement of all past and future disputes. Iraq viewed such an arbitration as inconsistent with the arbitration provisions included in the concession agreements. "Its purpose was not to provide for an orderly way of settling future disputes relating to royalty expensing, but to have foreign private persons sit in judgment on the validity of Law No. 80 in Iraq."[72] The companies' refusal to expense royalty deprived Iraq, until the Tehran Agreement was signed on February 14, 1971, of about ID 100 million.[73]

The companies came close to rendering Law No. 80 inoperative. It appears that during the second round of negotiations, which lasted from February 5 to June 3, 1965, the companies were successful in convincing the government negotiator that the stagnation that characterized the oil industry in Iraq since the promulgation of Law No. 80 could not be removed until and unless the dispute concerning that law was resolved to the companies' satisfaction.[74] Accordingly, on June 3, 1965, two draft agreements were initialed. The first agreement restored to the companies areas totaling 1,937 square kilometers. This was permissible under article 3 of Law No. 80. The areas in question contained all of Iraq's proved oil reserves, including the very rich Rumaila area in the southeastern part of the country.[75] The second agreement established the Baghdad Petroleum Company as a joint venture between INOC and the IPC group with the former holding one-third of the shares. The newly created company was granted a concession to operate in an area measuring 320 square kilometers.[76] The proposed deal ran into stiff public opposition and, consequently, was never ratified.[77]

Nationalization

Between 1966 and 1971 the list of complaints grew longer, while the possibility of reconciling the differences that separated the government from the oil companies became dimmer. In addition to its long standing demands, the government added the following: (1) production programming, (2) increasing the companies' investment expenditures, (3) bringing the price of southern oil to parity with other Arabian Gulf crudes, (4) moving IPC's headquarters to Baghdad, and (5) allowing the government some degree of financial control over the operations of the oil companies.[78] The last demand is significant. The government's grievances were no longer confined to questions of accounting procedures, but, rather, the government was demanding a measure of control over expenditures.

To press its demands, the government adopted certain measures that effectively ended any hopes that the companies may have had regarding the abrogation of Law No. 80. In 1967 the government promulgated a law prohibiting INOC from granting any concessions.[79] More significant was the decision by the government to

ROLE OF THE OIL SECTOR / 21

have INOC develop the North Rumaila Field in southern Iraq.[80] The oil companies had entertained hopes of recovering this "highly prolific field, with output per well around 11-2TBD."[81] Their hope was kept alive by article 3 of Law No. 80, which permitted the government to return to the companies additional areas other than those assigned to them under Law No. 80. "IPC was particularly anxious to recover North Rumaila; negotiations in 1965 nearly restored it, but eventually broke down—an incident well worth research."[82]

The loophole was plugged finally in 1970 when the government passed a law abrogating article 3 of Law No. 80.[83] It will be explained shortly that the loss of the North Rumaila Fields prompted the companies to demand, during subsequent negotiations, an enormous compensation for what they termed the "economic consequences" of Law No. 80.

In the spring of 1971, the IPC group asked the government to initiate talks for the purpose of settling all outstanding issues between the two parties. The government agreed but made it clear that it did not consider Law No. 80 negotiable. The talks commenced on January 15, 1972, with the companies' insistence on solving all issues simultaneously in a package deal that included the economic consequences of Law No. 80. They offered to expense royalty retroactively on northern crude oil on the basis of border value rather than posted prices, to increase production from Basrah, and to pay a lump sum of £10 million as compensation covering all the government's outstanding financial claims (estimated by the Ministry of Oil and Minerals to exceed £650 million). In exchange the companies made the following demands: (1) a 20-year contract enabling them to purchase 8 billion barrels of Rumaila crude produced by INOC at $1.62 per barrel plus a yearly increment, as provided by the Tehran Agreement and (2) the surrender by INOC to the companies free of charge of 12.5 percent of INOC's production, over and above the quantities to be purchased by them. This extraordinary demand was made because the companies felt that they were entitled to receive royalty as a form of compensation for Law No. 80.[84]

The government rejected the companies' demands for two reasons. First, it already had informed them that it would not consider any suggestion affecting Law No. 80 in any way. Second, accession to their demands would leave INOC without any marketing opportunities free from the companies' influence. The 8 billion barrels they asked for represented all the recoverable reserves of North Rumaila. Furthermore, the price they offered for the crude "would, in fact, be lower than the companies would pay were the companies themselves to undertake the development of North Rumaila Fields."[85]

The government offered to sell the companies 150 million barrels of North Rumaila crude over a period of ten years at commercial prices and conditions. Coupled with this offer was a reminder to the companies that the government categorically rejected the principle of compensating the companies for Law No. 80, as such a principle would be an infringement upon the country's sovereignty, that it preferred to deal with each issue individually and separately, and that it insisted on production programming and being compensated for losses resulting from the policy of stagnant production adopted by the companies ever since Law No. 80 had been promulgated.[86]

The government invited the companies to submit a new and positive offer that could be used as a basis for negotiations. No such offer was made. Instead, the companies resorted to their old weapon of reduced production. The end of February 1972 witnessed a sudden and drastic reduction in production from the northern fields after such production had reached its maximum of 4.75 million tons per month during December, January, and early February. In March production came down to 3.39 million tons and declined in April and May to 2.5 million tons per month.[87]

The new production policy spelled financial disaster to the country. Were production allowed to continue at this low rate, the total drop in government earnings for the year 1972 would have exceeded £110 million. This drop in revenues represented more than 50 percent of the total investment program for the fiscal year 1972/73.[88] Hence, depriving the country of these sums would have resulted in paralyzing the progress of Iraq's economic development.

Reduced production was justified on the grounds of reduced demand for oil in Western Europe and Japan. To accommodate the companies, the government offered three alternatives:

1. The companies may produce at full capacity of 57 million tons annually (MTA), off-take 30 MTA (the amount they claim there is a demand for), and leave the remaining 27 MTA for the government without obligation, apart from cost of production and transportation. The government would be free to market the oil.

2. If the companies believed they were unable, within their commercial interests, to produce and export more than 30 MTA, then they may continue to do so provided that they agree to surrender the excess capacity of existing installations, especially the pipelines to the government to use as it deemed fit. All installations erected in Iraq are in fact financed jointly by both parties with Iraq paying its share in the

form of amortized costs. It was not in Iraq's interest that a portion of this capacity was left idle.

3. If, as the companies claimed, production from the Northern fields was less profitable than production from the Southern fields, then they are invited to hand over the Northern fields to the government. They would be amply compensated by increasing production from the South since the Southern fields under their control contained sufficient oil reserves to raise the production rate to a level that would make up for the production from the North.[89]

As there was no response from the companies to these proposals, the Revolutionary Command Council issued on May 17, 1972, an ultimatum demanding the restoration of production rates to peak capacity and the submission of a positive offer concerning other government demands. The companies were given two weeks to reply. On May 31, 1972, the companies submitted a last offer, which in substance was very much like the previous offers. Their offer was unacceptable and on June 1, 1972, the government promulgated Law No. 69, which provided for the nationalization of the operations of IPC but left unaffected IPC's two affiliates, MPC and BPC.[90]

Article 3 of the law promised IPC full compensation, and the company was invited immediately to enter negotiation for this purpose. With the help of OPEC's secretary general, the two parties were able to conduct meaningful talks. As a result, on February 28, 1973, an agreement was signed in Baghdad between the government and the IPC group in which all outstanding issues were settled.[91] The agreement provided for the transfer of the assets and concessions of both IPC and MPC to Iraq but left the BPC concession intact. In addition, the companies agreed to pay Iraq £ 141 million in settlement of all outstanding claims by the government and to effect a substantial increase in production from BPC's fields with a view toward reaching an annual average of 80 million tons by 1976. In return, the government agreed to compensate IPC for its nationalized assets through the delivery of 15 million tons (approximately 105 million barrels) of Kirkuk crude, free on board (f.o.b.), Mediterranean ports, at an average rate of 1 million tons per month.

During and immediately following the October 1973 Arab-Israeli War, the government passed three laws providing for the nationalization of U.S., Dutch, and Gulbenkian interests in BPC.[92] Two years later (that is, on December 8, 1975) the remaining foreign interests in BPC (British and French) were nationalized.[93] The nationalization law made INOC fully responsible for operating the facilities of BPC. As in the case of IPC, the laws nationalizing BPC contained provisions for compensation.

24 / OIL REVENUES & ACCELERATED GROWTH

The toughest postnationalization problem was that of marketing. Following nationalization on June 1, 1972, IPC applied pressure, including the threat of legal action, to prevent third parties from purchasing Iraqi oil.[94] Fortunately, in 1972 the government was able to conclude sales agreements with a number of companies and countries that helped to limit the decline in exports during that year. The most significant sales agreement both politically and economically was the one concluded with France on June 18, 1972, during the visit of the vice chairman of the Revolutionary Command Council to France.[95] Under the terms of that agreement, Iraq agreed to supply Compagnie Française de Pétroles (CFP), a shareholder of the nationalized IPC, a quantity of crude oil from the nationalized Kirkuk fields equal to 23.75 percent of production—that is, equivalent to the CFP's shareholding interest in IPC.[96] Payment arrangements under this agreement were on the basis of the financial terms prevailing prior to nationalization at a price equal to the tax-paid cost and subject to increases in posted prices and other financial adjustments agreed upon within the framework of OPEC.[97]

Coming as it did, only 18 days after nationalization, the agreement with France was a real victory, particularly in view of the fact that CFP was a shareholder of IPC. It frustrated IPC's efforts to block the marketing of Iraqi oil. After that, INOC was able to sign contracts to sell Iraqi crude to the Soviet Union, Bulgaria, German Democratic Republic (GDR), India, Italy, Spain, Tunisia, Egypt, Austria, Belgium, Yugoslavia, Turkey, Finland, Morocco, and others.[98] In brief, the battle of nationalization was won. Iraq's success in marketing nationalized oil, especially the agreement with France, convinced the IPC group of the impossibility of reversing the nationalization and thus paved the way for the February 28, 1973 accord, which marked the end of a half a century of foreign domination over the most vital sector in the Iraqi economy.

PRODUCTION, REFINING, AND EXPORTS

Production

Table 2.1 shows Iraq's crude oil production and export for the 1950-75 period. Although production in 1975 was almost 17 times the production in 1950, the pattern of growth was erratic throughout this period. After a high rate of growth, 22 percent per annum during the 1950s, expansion of output decelerated to a rate below 5 percent during the 1960s; it recovered only slightly in the early 1970s (Table 2.2). The relatively low rate of growth during

TABLE 2.1

Iraq: Crude Petroleum Production and Export
(millions of metric tons)

Year	1 Production	2 Export	3 2 as Percent of 1
1950	6.6	6.1	92.4
1951	8.6	8.0	93.0
1952	18.5	17.9	96.8
1953	28.2	27.4	97.2
1954	30.6	29.6	96.7
1955	33.2	32.2	97.0
1956	31.5	29.6	94.0
1957	22.0	20.2	91.8
1958	35.8	33.8	94.4
1959	41.9	39.8	95.0
1960	47.5	45.2	95.2
1961	49.0	46.6	95.1
1962	49.2	46.7	94.9
1963	56.7	54.1	95.4
1964	61.6	59.2	96.1
1965	64.5	61.6	95.5
1966	68.0	62.8	92.4
1967	59.9	57.4	95.8
1968	73.8	70.4	95.4
1969	74.5	70.8	95.0
1970	76.4	73.3	95.9
1971	83.1	79.4	95.5
1972	71.1	68.4	96.2
1973	99.4	94.7	95.3
1974	96.9	90.0	92.9
1975	111.2	103.2	92.8

Sources: United Nations, Department of Economic and Social Affairs, Statistical Office, World Energy Supplies: 1950-1974 (ST/ESA/STAT/SER. J/19); and United Nations, Department of Economic and Social Affairs, Statistical Office, World Energy Supplies: 1971-1975 (ST/ESA/STAT/SER. J/20).

TABLE 2.2

Crude Petroleum Production in Selected Countries
(millions of metric tons)

	Year					Rates of Growth (percent)		
	1950	1960	1970	1975	1950-60	1960-70	1970-75
Iran	32.2	53.5	191.3	267.6	5.2	13.6	6.9
Iraq	6.6	47.5	76.4	111.2	22.0	4.9	7.8
Kuwait	17.2	84.8	150.6	105.2	17.3	5.9	-6.9
Saudi Arabia	26.6	64.5	188.4	352.4	9.3	11.3	13.4
Middle East	85.4	261.3	694.5	975.6	11.8	10.3	7.1
World	520.4	1,052.1	2,270.2	2,646.8	7.3	8.0	3.1
Iraq's share in (percent)							
Middle East production	7.7	18.2	11.0	11.4			
World production	1.3	4.5	3.4	4.2			

Sources: United Nations, Department of Economic and Social Affairs, Statistical Office, World Energy Supplies: 1950-1974 (ST/ESA/STAT/SER. J/19); and United Nations, Department of Economic and Social Affairs, Statistical Office, World Energy Supplies: 1971-1975 (ST/ESA/STAT/SER. J/20).

the 1960s and early 1970s was due primarily to sluggish investment in the oil sector; it also reflected the practice, often followed by oil companies, of reducing production as a means of pressuring the Iraqi government to moderate its demands.

The low rate of growth in Iraq's crude oil production in the 1960s stands out when compared with growth rates in other major producing countries in the area. As shown in Table 2.2, the 4.9 percent rate of growth in Iraq's production in the 1960-70 period was significantly lower than the 13.6 percent growth rate in Iran and the 11.3 percent rate experienced by Saudi Arabia. Only in Kuwait was the rate of growth close to that of Iraq. However, this was the result of a deliberate policy of oil conservation by the Kuwaiti authorities.

Examining the growth pattern in the present decade, crude oil production rose 9.7 percent in 1971 largely because of the improved relations between the government and the foreign oil companies following the signing of the Tehran and East Mediterranean agreements.[99] However, it declined by 15.2 percent in 1972, owing to IPC's decision to reduce production from the Kirkuk oil fields during the first five months of the year and a reduction in output during the balance of the year because of the initial impact of nationalization. Output rose sharply in 1973 (39.8 percent) as a result of the restoration to normal production of the nationalized oil fields and the continued growth of output from the southern fields. In 1974 oil production experienced a modest decline (-2.5 percent) but made a healthy recovery in 1975. In short, after the impressive rate of the 1950s, Iraq's output has been subject to fluctuations since 1960. On the other hand, the output of neighboring countries exhibited a steady and significant rate of growth.

Refining

Seven refineries were operating in Iraq in 1974 with a total capacity of 183,500 barrels per day.[100] The Daurah and Basrah refineries are the largest—each with a capacity of 70,000 barrels per day. The Kirkuk refinery is next in size with a capacity of 20,000 barrels per day. The other five refineries are very small with a total combined capacity of 20,300 barrels per day.[101]

Iraq's refinery capacity is quite meager for a major oil producer with a productive capacity of nearly 3 million barrels per day. This anomalous situation was created by the foreign oil companies that had restricted their activities in Iraq to exploration and production only. In the Arabian Gulf region, Iraq was the only major producer where the operating companies failed to establish an export refinery.[102] The few small refineries that were built were intended to meet local requirements of the oil companies.[103] In fact, it was not until the 1950s that the government was able to build a refinery of its own (Daurah refinery near Baghdad) to meet part of the country's demand for refined products.[104]

Since 1973 the country has been implementing several projects designed to further expand its refinery capacity. The output of some of the new refineries is intended for external markets.

Exports

Iraq exports most of its crude oil output; domestic consumption accounts for only about 5 percent of total production (Table 2.1). The principal market for Iraqi oil exports is Western Europe. Western European countries accounted for 75 percent of Iraq's total crude oil exports in 1967, 75.9 percent in 1971, and 54.1 percent in 1975.[105] The relative decline in Iraq's oil exports to Western Europe in 1975 reflected the sharp rise in the country's exports to the socialist countries.

Among Western European countries, France and Italy are the principal importers of Iraqi crude, accounting for about 55 percent of all crude oil shipped from Iraq to Western Europe in 1975.[106]

THE IMPACT OF THE OIL SECTOR

The oil sector has been the dominant one in the Iraqi economy since the early 1950s. It accounted for more than one third of the country's gross domestic product (GDP) during the 1953-73 period. Following the rise in oil prices, the share of the oil sector in Iraq's GDP rose sharply, amounting to 60 percent in 1974, 57 percent in 1975, and 53 percent in 1976 (Table D.2).

In contrast to the rising share of oil in total output, that of agriculture has declined rather sharply since the early 1950s. The relative importance of the other sectors has varied considerably over the years, though the trend has generally been upward (Table 2.3).

The diminishing role of agriculture and the growing dominance of oil were not accompanied by similar changes in the pattern of employment. For example, agricultural employment as a percentage of Iraq's total employment increased from about 49 percent in 1963 to 56 percent in 1973 (Table 2.4). During that same period, the oil industry's share of total employment declined moderately.

The oil industry is capital intensive, and, therefore, it cannot absorb a substantial percentage of the labor force. However, this technical attribute was reinforced by the "foreign enclave" nature of the oil industry in Iraq. As mentioned earlier, the foreign oil companies confined their operations to extraction and export; they refrained from investing in downstream operations such as refining. This fact together with the inability of the government to establish petrochemical and other oil-based industries deprived the country of the benefits that would have accrued as a result of linkages.

TABLE 2.3

Iraq: Sectoral Contribution to Gross Domestic Product
(in percent at current prices)

Sector	1953	1963	1973	1976
Agriculture, forestry, and fishing	22.1	16.3	14.2	7.6
Oil extraction	39.9	36.2	35.5	53.4
Other mining and quarrying	0.3	0.3	0.7	0.6
Manufacturing	6.1	9.6	9.9	7.1
Construction	3.5	3.0	3.6	7.7
Electricity, water, and gas	0.5	0.8	1.0	0.5
Transport, communication, and storage	6.6	7.3	5.6	4.8
Wholesale and retail trade	5.5	5.4	7.3	4.3
Banking, Insurance, and real estate	1.0	1.9	1.3	1.9
Ownership of dwellings	3.6	1.8	3.7	1.7
Public administration and defense	5.7	10.0	9.7	9.4
Other services	5.2	7.4	7.5	1.0
GDP at factor cost	100.0	100.0	100.0	100.0

Source: Derived from Table D.2.

The oil industry, in other words, failed to play the role of leading sector. However, its fiscal function continues to grow in importance. Oil revenues are the major source of government finance, development finance, and foreign exchange. The rest of the chapter examines these aspects in some detail.

Impact on the Balance of Payments

The contribution of the oil sector to Iraq's balance of payments is very substantial. Net foreign exchange receipts generated by the

TABLE 2.4

Iraq: The Structure of Employment
(thousands)

Sector	1963 Number	1963 Percent of Total	1968 Number	1968 Percent of Total	1973 Number	1973 Percent of Total
Agriculture	873.5	48.88	1,253.6	53.96	1,540.4	55.77
Mining (oil)	12.5	0.70	15.0	0.65	18.5	0.67
Manufacturing	130.0	7.27	146.0	6.28	170.0	6.15
Electricity, gas, and water	12.0	0.67	12.8	0.55	14.3	0.52
Construction	43.1	2.41	66.0	2.85	73.0	2.64
Commerce	115.0	6.44	140.0	6.02	164.0	5.94
Transport	121.0	6.77	140.0	6.02	162.0	5.86
Services	260.0	14.55	290.0	12.48	330.0	11.95
Other	220.0	12.31	260.0	11.19	290.0	10.50
Total	1,787.1	100.00	2,323.4	100.00	2,762.2	100.00

Sources: Iraq, CSO, Annual Abstract of Statistics 1972; and Iraq, CSO, Annual Abstract of Statistics 1973.

oil sector amounted to ID 199.4 million in 1970, ID 258.4 million in 1973, and ID 1,469.2 million in 1974.[107] In fact, net foreign exchange receipts from the oil sector have financed the nonoil current account deficits since 1950.

Table D.1 highlights Iraq's current account position for the years from 1950 to 1974. Oil exports during this period constituted, on the average, about 80 percent of the country's total exports. Its contribution grew markedly, from 47.5 percent in 1950 to 98.3 percent in 1974. There is no doubt that oil exports will dominate Iraq's merchandise trade and current accounts in the foreseeable future.

Gross foreign exchange generated from oil exports totaled ID 7,950 million in the 1950-73 period; of this amount 47.4 percent (ID 3,770 million) was used for repatriating the profits of the foreign oil companies. The remaining 52.6 percent plus the foreign exchange Iraq was able to earn from its other exports, amounting together to ID 5,621 million, was available for meeting the cost of goods and services imported by the country. Out of this amount only ID 4,202 million or 74.6 percent was actually used during this period.

This favorable financial position enabled Iraq to import both consumer and capital goods without having to rely on foreign loans or economic assistance from other countries. In fact, the country always had enjoyed merchandise trade surpluses, the magnitude of which increased from ID 11.4 million in 1950 to ID 1,071.7 million in 1974. The balance on current account had shifted from a small deficit of ID 1.5 million in 1950 to a sizable surplus of ID 690.8 million in 1974. Between 1950 and 1974 Iraq displayed a deficit on current account only in six years. Two of these deficits, those incurred in 1956 and 1957, were the result of the Suez War of 1956. The deficits of 1960 and 1961 were due to the strained relations between the government and the foreign oil companies that adversely affected oil exports.

The performance of Iraq is out of character for a developing country where the normal situation is to finance current account deficits with capital inflows. Indeed, the rise of oil production transformed Iraq into an export economy. In 1974, the ratio of exports to GDP amounted to more than 61 percent, whereas the ratio of imports was less than 32 percent.[108]

Oil Revenues and Public Finance

In the early years oil revenues were too small to play a major role in government finance. For example, they amounted to approximately ID 0.4 million in 1931, ID 1.6 million in 1941, and ID 6.7

million in 1950.[109] However, following the profit-sharing agreement of 1952, oil revenues rose sharply, reaching ID 40.1 million in 1952, ID 58.3 million in 1953, ID 64.3 million in 1954, and ID 73.7 million in 1955 (Table 2.5). Except for a brief decline in 1956 and 1957 caused by the interruption of the flow of oil through Syria during the Suez conflict, oil revenues accruing to the government continued to show substantial increases reaching a level of ID 1,724.1 million in 1974. As a share of total government revenues, oil receipts reached a high of about 84 percent in fiscal year 1974, compared to 31 percent in fiscal 1951.*

Oil Revenues and Economic Development

In 1950 Iraq established a Development Board (DB) and assigned to it the task of preparing a general plan for developing the resources of the country, undertaking the execution of projects and turning the completed projects to the ministries concerned with administration and maintenance.[110] Initially, all revenues from oil were allocated to the DB, but these allocations were later reduced to 70 percent in 1952 and finally to 50 percent in 1959.[111]

During the 1951-74 period, six development programs and plans were prepared, and every one of them relied almost exclusively on oil revenues for financing. As shown in Table 2.6, during the early years oil revenues were projected to constitute, on the average, almost 98 percent of the total revenues needed to finance development programs. Even after the government decided to reduce the share of the development budget in oil revenues to 50 percent, these revenues continued to be the primary source of finances for development. Between 1959 and 1974, oil revenues were expected to finance 75.2 percent of total development expenditures. In fact, out of ID 2,635.8 million actually realized and earmarked to finance the investment expenditures of the central government during the period 1951-74, ID 2,390.2 million, or 90 percent, came out of oil revenues.

Thus, from the very beginning planners in Iraq were assured of a steady flow of capital funds. This favorable position enabled the country to undertake ambitious development programs without having

---

*The fiscal year in Iraq starts on April 1 and ends on March 31. Thus, for example, fiscal year 1973 refers to the period from April 1, 1973 to March 31, 1974. All years in this study refer to fiscal years unless otherwise indicated. Effective January 1, 1976, the fiscal year was changed to coincide with the calendar year.

TABLE 2.5

Iraq: Proportion of Oil Revenues to
Total Government Revenues
(millions of Iraqi dinars)

Year	1 Oil Revenues	2 Total Revenues	3 1 to 2 (percent)
1951	13.9	44.9	31.0
1952	40.1	74.4	53.9
1953	58.3	82.9	70.3
1954	64.3	97.8	65.7
1955	73.7	125.9	58.5
1956	68.8	113.8	60.4
1957	48.8	97.6	50.0
1958	79.8	137.2	58.2
1959	86.6	133.2	65.0
1960	95.1	151.2	62.9
1961	94.1	187.7	50.1
1962	95.1	184.7	51.5
1963	110.0	194.3	56.6
1964	126.0	221.0	57.0
1965	134.0	254.0	52.8
1966	122.4	229.4	53.4
1967	151.7	292.2	51.9
1968	174.7	308.9	56.6
1969	169.0	341.4	49.5
1970	211.7	403.7	52.4
1971	350.0	534.1	65.5
1972	218.6	406.4	53.8
1973	823.2	1,039.7	79.2
1974	1,724.1	2,062.0	83.6

Sources: Ferhang Jalal, The Role of Government in the Industrialization of Iraq (London: Frank Cass, 1972), p. 11; Iraq, CBI, Bulletin, no. 3 (July-September 1971); Iraq, CBI, Bulletin, no. 1 (January-March 1976); and Iraq, CSO, Annual Abstract of Statistics 1976.

TABLE 2.6

Iraq: Revenues of Economic Development Programs and Plans, 1951-74
(millions of Iraqi dinars)

	Estimated Revenues			Actual Revenues		
	1 Total	2 Oil	3 2 to 1 (percent)	1 Total	2 Oil	3 2 to 1 (percent)
Revised First General Program (1951-54)	99.9	96.0	96.1	107.5	104.4	97.1
Revised Second General Program (1955-59)	324.2	319.9	98.7	241.4	234.1	97.0
Provisional Economic Plan (PEP) (1959-61)*	100.9	94.1	93.3	100.9	94.1	93.3
Detailed Economic Plan (1961-64)	454.1	252.6	55.6	239.0	195.6	81.8
Five-Year Economic Plan (1965-69)	561.2	390.0	69.5	407.0	372.3	91.5
National Development Plan (1970-74)	1,932.0	1,554.4	80.4	1,540.0	1,389.7	90.2
Total	3,472.3	2,707.0	78.0	2,635.8	2,390.2	90.7

*Since the PEP did not provide any revenue estimates, it is assumed here that estimated revenues were equal to actual revenues. The planner was satisfied to state only that the plan would be financed from the following sources: 50 percent of oil revenues, other revenues, and foreign loans. See Republic of Iraq, Law No. 181 of 1959.

Sources: Iraq, CSO, Annual Abstract of Statistics 1976; and Iraq, Ministry of Planning, Results of the Follow-Up of the Implementation of Investment Objectives of Economic Programs and Plans in Iraq: 1951-1971 (Baghdad, 1972), pp. 10, 20, 46, 60, 82 (hereafter cited as Economic Plans in Iraq).

to squeeze domestic consumption too tightly to provide the needed capital funds or to resort to excessive foreign borrowing to provide the needed foreign exchange.

Indeed, Iraq could afford to undertake heavy development expenditures and at the same time avoid the threat of balance of payment difficulties so common on many developing countries. However, despite these advantageous conditions, absorptive capacity constraints caused the country's accomplishments to lag far behind its expectations. More will appear on this point in the following three chapters.

CONCLUSION

In addition to providing a historical survey of developments in the oil field, this chapter highlighted the role of the oil sector in the Iraqi economy. Whether judged by its contribution to gross national product, foreign exchange earnings, government revenues, or development finance, the preeminence of this sector cannot be doubted. However, the oil sector remained an enclave modern sector in an otherwise backward economy.

The link between the oil industry and the rest of the economy is very weak. As indicated earlier, the oil sector employs less than two-thirds of 1 percent of the labor force. As such, its direct contribution to incomes and aggregate demand is minor. Nor has it ever been in a position to secrete "abilities, skills, and attitudes" needed for further development of the economy.

Furthermore, owing to the general backwardness of the economy, the failure of the foreign oil companies to develop downstream operations (such as refining), and the failure of the government to build petrochemical industries, the Iraqi economy was deprived of the beneficial effects of the backward, lateral, and forward linkages that normally accrue to an economy characterized by the ascendancy of a leading sector.

NOTES

1. According to the census conducted on October 17, 1977, Iraq's population is 12,171,480. See The Middle East, December 1977, p. 21.

2. Abdul Jalil el-Hadithy and Ahmad el-Dujaili, "Problems of Implementation of Agrarian Reform in Iraq," in Land Policy in the Near East, ed. Mohamad Riad el-Ghomeny (Rome, Italy: Food and Agriculture Organization of the United Nations, 1967), p. 219.

36 / OIL REVENUES & ACCELERATED GROWTH

 3. Lord (James Arthur) Salter, The Development of Iraq: A Plan of Action (Baghdad: Iraq Development Board, 1955), p. 157. According to Doreen Warriner, per head agricultural land in Iraq is "somewhat more than in Syria and twenty times as much as in Egypt." See Doreen Warriner, Land Reform and Development in the Middle East, 2d ed. (London: Oxford University Press, 1962), p. 115.
 4. Oil and Gas Journal, February 26, 1979, p. 166.
 5. Stephen H. Longrigg, Oil in the Middle East, 3d ed. (London: Oxford University Press, 1968), p. 10.
 6. Benjamin Shwadran, The Middle East, Oil and the Great Powers, 2d ed., rev. (New York: Council for Middle Eastern Affairs Press, 1959), p. 193.
 7. Ibid.
 8. Ibid.
 9. Ibid.
 10. Ibid., p. 194. The U.S. group was represented by Rear Admiral Colby M. Chester, who had the backing of the New York Chamber of Commerce and the New York State Board of Trade and the support of President Theodore Roosevelt and Secretary of State Elihu Root.
 11. Ibid.
 12. Ibid.
 13. Ibid
 14. Ibid.
 15. Ibid.
 16. Ibid., p. 196.
 17. For the text of the grand vizier's letter, see Shwadran, Middle East, Oil and the Great Powers, p. 196.
 18. Longrigg, Oil in the Middle East, p. 44.
 19. Ibid., pp. 68-69.
 20. Iraq, Ministry of Oil and Minerals, The Nationalization of Iraq Petroleum Company's Operations in Iraq: The Facts and the Causes (Baghdad, Iraq, 1973), p. 3 (hereafter cited as Nationalization of IPC).
 21. Shwadran, Middle East, Oil and the Great Powers, pp. 205-16. The position taken by the State Department was not inspired by any desire to safeguard the interest of the Iraqi people; rather, it was part of its campaign to secure a share for U.S. oil companies in TPC.
 22. Ibid., p. 214.
 23. Nationalization of IPC, p. 3.
 24. Ibid.
 25. Muhsen al-Mosawi, Iraq's Oil: The People's Struggle against Oil Companies' Covets (Baghdad, Iraq: Ministry of Information, 1973), p. 42.

26. Ibid. The constitution was ratified by the Constituent Assembly three days after the concession agreement was signed.
27. Nationalization of IPC, p. 4.
28. Ibid.
29. Shwadran, Middle East, Oil and the Great Powers, p. 197.
30. Ibid., p. 198.
31. Ibid.
32. Ibid., pp. 205-6.
33. Longrigg, Oil in the Middle East, p. 70.
34. Ibid.
35. Ibid., pp. 69-70.
36. Shwadran, Middle East, Oil and the Great Powers, p. 247, fn. 10.
37. Nationalization of IPC, p. 4.
38. Longrigg, Oil in the Middle East, p. 70.
39. Ibid., p. 74.
40. Ibid.
41. Ibid., p. 80.
42. Shwadran, Middle East, Oil and the Great Powers, p. 250.
43. Longrigg, Oil in the Middle East, p. 82, fn. 1.
44. Ibid., p. 190.
45. Ibid., p. 189.
46. Zuhayr Mikdashi, A Financial Analysis of Middle Eastern Oil Concessions: 1901-65 (New York: Praeger, 1966), p. 68.
47. Ibid., p. 206.
48. Ibid.
49. Ibid.
50. Longrigg, Oil in the Middle East, p. 190.
51. Shwadran, Middle East, Oil and the Great Powers, pp. 106-7, 390.
52. al-Mosawi, Iraq's Oil, p. 68.
53. Longrigg, Oil in the Middle East, p. 191.
54. Ibid.
55. Ibid.
56. Ibid., pp. 191-92.
57. Mikdashi, Financial Analysis of Middle Eastern Oil Concessions, pp. 197-99.
58. Ibid., p. 199.
59. Ibid.
60. Ibid., p. 201.
61. Ibid., p. 200.
62. Ibid.
63. Ibid.
64. Ibid., p. 206.
65. Ibid., pp. 206-7.

66. Ibid., p. 207.
67. <u>Nationalization of IPC</u>, p. 6.
68. Republic of Iraq, <u>Law No. 80 of 1961</u>.
69. Republic of Iraq, <u>Law No. 11 of 1964</u>.
70. <u>Nationalization of IPC</u>, p. 8.
71. Figures are derived from table 4 (p. 7) in Jawad M. Hashim, "The Structure of Fixed Capital Formation by Foreign Oil Companies Operating in Iraq During 1957-1969 and Its Contribution to the Iraqi Economy" (Paper presented before the Eighth Arab Petroleum Congress, Algiers, May 28-June 3, 1972). In his introduction the author states that, "Gross fixed capital formation in oil companies is defined to comprise annual expenditure on replacement of, and additions and major improvements to, fixed capital assets including the value of work-in-progress, within the political boundaries of Iraq." Clearly, the companies were actually disinvesting because depreciation allowances must have exceeded gross investment expenditures manyfold. It is not difficult, therefore, to recognize that the companies were engaged deliberately in the destruction of the Iraqi economy.
72. <u>Nationalization of IPC</u>, p. 12.
73. Ibid., p. 10.
74. al-Mosawi, <u>Iraq's Oil</u>, pp. 110-11.
75. Ibid., p. 107.
76. Ibid.
77. Ibid., p. 113.
78. Ibid., pp. 128-48.
79. Republic of Iraq, <u>Law No. 97 of 1967</u>.
80. M. A. Adelman, <u>The World Petroleum Market</u> (Baltimore: Johns Hopkins University Press, 1972), p. 223.
81. Ibid.
82. Ibid.
83. Republic of Iraq, <u>Law No. 24 of 1970</u>.
84. <u>Nationalization of IPC</u>, pp. 12-16.
85. Ibid., p. 15.
86. Ibid., pp. 16-17.
87. Ibid.
88. Ibid., p. 19.
89. Ibid., pp. 20-21.
90. Republic of Iraq, <u>Law No. 69 of 1972</u>.
91. For the text of the Heads of Agreements see <u>Nationalization of IPC</u>, app. C, pp. 33-35.
92. <u>Law No. 70 of 1973</u> nationalized the interests of Exxon and Mobile (23.75 percent); <u>Law No. 90 of 1973</u> nationalized the interest of Royal Dutch Oil Company (14.25 percent); and <u>Law No. 101 of 1973</u> nationalized the interest of Partex (5 percent). These laws

were promulgated on October 7, October 21, and December 29, 1973, respectively.

93. Republic of Iraq, Law No. 200 of 1975.

94. For a survey of the specific actions taken by IPC to prevent the marketing of Iraqi oil see Al-Naft Wal-Aalam, June 1973, pp. 17-18.

95. Al-Naft Wal-Aalam, March 1973, p. 35.

96. Ibid.

97. Ibid.

98. Iraq, INOC, Annual Review 1972, p. 23.

99. After a decade of stagnation, the prices of crude oil began to rise late in 1970. The first breakthrough was accomplished by Libya, which, in September 1970, concluded a series of agreements with the oil companies that resulted in an increase of $0.30 per barrel in the posted price of its crude oil and raised the tax rate from 50 percent to an average of 55 percent. Effective November 14, 1970, tax rates of crude oil exported from the Arabian Gulf were raised similarly and posted prices of some medium and heavy crudes were increased a few cents per barrel.

These developments were soon followed by two major agreements: the Tehran Agreement and the Tripoli Agreement. The former was signed in Tehran on February 15, 1971, by six Arabian Gulf states and 13 oil companies operating in these states. The latter, which became effective on March 20, 1971, was concluded between the Libyan government and the oil companies and served as a model for the East Mediterranean Agreements concluded during June 1971 between Iraq and Saudi Arabia, on the one hand, and the oil companies, on the other. As a result of these agreements, posted prices were increased by $0.35 per barrel at the Arabian Gulf terminals and $0.42 per barrel at the Mediterranean terminals. See the following issues of the Middle East Economic Survey: September 11, 1970; September 25, 1970; October 2, 1970; November 20, 1970; November 27, 1970; February 19, 1971; and April 2, 1971.

100. OPEC, Annual Statistical Bulletin 1974 (Vienna, 1975), p. 45.

101. Ibid.

102. Nationalization of IPC, p. 10.

103. Ibid.

104. Ibid.

105. United Nations, Department of Economic and Social Affairs, Statistical Office, World Energy Supplies: 1961-1970 (ST/STAT/SER. J/15); and World Energy Supplies 1971-1975 (ST/ESA/STAT/SER. J/20).

106. Ibid.

107. Iraq, CBI, Bulletin, no. 1, January-March 1976, pp. 70-71.

108. IMF, International Financial Statistics, April 1978, p. 161.
109. OPEC, Annual Statistical Bulletin 1974, p. 147.
110. Abbas Alnasrawi, Financing Economic Development in Iraq (New York: Praeger, 1967), p. 36.
111. Ibid., pp. 37, 42.

3

DEVELOPMENT PLANNING IN IRAQ: SURVEY AND EVALUATION

SURVEY OF DEVELOPMENT PLANNING

Since the establishment of the DB in 1950, development planning in Iraq has passed through through three distinct phases. The first phase covered the era of the DB. Under the DB no economic development plans as such were formulated. Rather, the development effort consisted of investment programs that were merely lists of public investment projects. Four such investment programs were drawn up between 1951 and 1956. The common feature that characterized all of them was their emphasis on irrigation, flood control, bridges, and harbors.

After the revolution of July 14, 1958, the DB was abolished and economic development policy entered its second phase, which can best be described as the era of partial national planning. The revolutionary regime was determined to effect a radical agrarian reform, accelerate the pace of industrialization, expand spending on social welfare projects (such as public housing, schools, and hospitals), and implement a new foreign trade policy.

The reformed development machinery comprising the Planning Board and the Ministry of Planning proceeded to formulate a new economic plan that was more consistent and efficient, as compared with the development programs of the DB. Nevertheless, economic planning during this phase was partial rather than comprehensive, because the planner failed to give due attention to social, educational, and manpower planning. Nor did he recognize explicitly the role of the private sector in the implementation of development plans.

The third phase commenced with the introduction of the Five-Year Economic Plan (1965-69). The new phase introduced no radical changes on the objectives of development, although planning had gradually become more comprehensive in scope. In preparing the five-year plan, the planner made a genuine effort to incorporate the areas that were overlooked in previous plans. The trend was broadened and carried out in a more formal fashion in the National Development Plan (NDP) (1970-74), which became operative on April 1, 1970.

The Development Board

Formal economic planning in Iraq was inaugurated in 1950 with the establishment of the DB. The law establishing the DB provided that all revenues from oil be placed at the DB's disposal.[1] The law also provided that the board be composed of five nonpolitical members and be placed under the chairmanship of the prime minister, with the minister of finance as ex officio member.

Endowed with ample financial resources, prestige, and power, the board was under pressure to commence operations.[2] In response, it produced a hurried five-year general program for the 1951-55 fiscal years. The program appropriated expenditures totaling ID 65.7 million to be financed out of an anticipated revenue of ID 95.1 million.[3]

The program was revised before implementation was started. The profit-sharing agreement that was concluded in early 1952 with the oil companies improved the financial position of the country. Guided by the recommendations of the World Bank Mission* and endowed with additional resources, the DB revised its first program by introducing a new six-year program for the years from 1951 to 1956. In addition to extending the period by another year, the revised version increased projected expenditures to ID 155.4 million. Revenues were estimated to total ID 168.7 million.[4]

To rectify certain shortcomings in the administrative structure of the development machinery, a law was passed providing for the creation of a Ministry of Development. The ministry was to perform the double function of serving as a link between the board and the Council of Ministers, on the one hand, and as the executive organ of the board, on the other.[5]

---

*At the request of the Iraqi government, the IBRD sent a special mission to Iraq for the purpose of surveying its resources and making appropriate recommendations. The mission was in Iraq between February and May 1951. Its report was submitted to the government in early 1952.

The new structure of the development machinery inaugurated its activities by replacing the existing program with a new five-year program for the years 1955-59. Once again both revenues and expenditures were revised upwards; they were estimated to be ID 215.7 million and ID 304.3 million, respectively. The projected deficit was to be financed by an anticipated rise in oil revenues during the years covered by the program.[6]

Even while drawing up this program, the DB was aware of the existence of certain fundamental problems that required solution. Prominent among these problems was the direction of the development effort. To help it resolve this and other difficulties the DB sought the advice of Lord (James Arthur) Salter as well as that of Arthur D. Little, Inc. Lord Salter's task, as stated in the preface to his report, was that of studying

> the timing and balance of the different projects of the Development Board; their co-ordination both with each other and with the action of other authorities; the impact of each upon others and upon the general economy of Iraq; and such modification in the administrative machinery both of the Board and of the Departments of the Government as is indeed to ensure their successful execution.[7]

Lord Salter submitted his "Plan of Action" in April 1955. The U.S. firm of Arthur D. Little, Inc. was commissioned to survey the industrial potential of the country. It submitted its report in May 1956.[8]

The completion of these studies and the availability of additional funds made possible by a further rise in oil revenues prompted the DB to revise its second general program. The law formalizing the revision extended the life of the program over another year and increased programmed expenditures to ID 500.1 million.[9] Revenues accruing to the revised program were estimated to total ID 390.1 million. The projected deficit of ID 110 million was to be financed by an anticipated rise in oil revenues as well as by an excess of allocations over actual expenditures.[10] That is to say, the DB was not sure that the investment program would be implemented as planned.

During the third year of the implementation of this program the revolution of July 14, 1958, occurred. A year later the program was suspended and replaced by the Provisional Economic Plan.

None of the programs adopted by the DB were ever completed. Changes in either economic or political circumstances invariably led to the replacement of an existing program by a new one. An unexpected rise in oil revenues or the completion of an economic study were the typical causes behind the revision and/or replacement of existing programs. On the other hand, political upheaval in 1958 was not only responsible for terminating the DB's last program but also for abolishing the DB itself.

Table 3.1 summarizes the planned and actual expenditures during the era of the DB. Note the low rates of implementation.* Also note the emphasis on agriculture and the low priority accorded industry.

TABLE 3.1

Iraq: Planned versus Actual Development
Expenditures, 1951-58
(millions of Iraqi dinars)

Sector	1 Planned	2 Actual	3 2 to 1 (percent)
Agriculture	178.2	81.3	45.6
Industry	71.2	34.8	48.9
Transport and communication	140.6	67.6	48.1
Building and services	108.1	81.7	75.6
Total	498.1	265.4	53.3

Source: Economic Plans in Iraq, pp. 8, 19.

Partial National Planning

Soon after it assumed power in 1958, the revolutionary government initiated a comprehensive review of development planning and development machinery. The review was concluded with the approval of the Executive Power Law, which abolished the Development Board and the Ministry of Development and established the Economic Planning Board and the Ministry of Planning.[11]

---

*Figures in this section as well as in subsequent sections dealing with completion rates refer to the totals of sectoral allocations and expenditures. Administration and follow-up allocations and expenditures are excluded.

The new structure inaugurated its activities by submitting the Provisional Economic Plan for the years 1959-62. The main purpose of the PEP was to provide an instrument for the completion of the projects that already had been started and to give the board and the Ministry of Planning sufficient time to draw up a detailed plan.[12] Total expenditures according to this interim plan were projected to be ID 392.1 million. There was no specific estimate of revenues except that financial resources were to be composed of 50 percent of oil revenues and the proceeds of two loans from the Soviet Union and Czechoslovakia amounting to ID 65 million and ID 12 million, respectively.[13]

Two years later the PEP was replaced by a five-year plan known as the Detailed Economic Plan (DEP) for the years 1961-65. Allocations under this plan totaled ID 556.3 million. The revenues of the plan were expected to come from the following sources: oil revenues, ID 315.8 million; foreign loans, ID 77.2 million; profits from public enterprises and revenue from miscellaneous sources, ID 30.8 million; and unidentified sources, ID 142.5 million.[14] In reality, the last item represented a gap between budgeted expenditures and revenues. In other words, the planner did not expect the plan to achieve its targets.

Officially, the DEP lasted only two years. On February 8, 1963, a new government assumed power. Transitional annual investment programs were drawn up for 1963 and 1964 that provided for programmed expenditures of ID 65.9 million and ID 106.8 million, respectively.[15]

As shown in Table 3.2, planned and actual allocations and expenditures during this period of partial national planning amounted to ID 756.0 million and ID 308.4 million, respectively. In addition to the very low rates of implementation, one also should note the shift in favor of industry and services. Furthermore, agricultural expenditures were shifted from flood control schemes to areas that directly contributed to increased agricultural production.

During this period the principle of centralized planning and decentralized implementation was instituted, and the Directorate of Central Statistics was attached to the Ministry of Planning.[16] In 1964 the Economic Planning Board was renamed the Planning Board and was reorganized so as to limit its membership to the prime minister; the ministers of planning, economics, and finance; the governor of the Central Bank; and four experts who do not hold any public office except their membership in the board. The law introducing the change also created a steering committee composed of the minister of planning as chairman and the four experts as members.[17]

TABLE 3.2

Iraq: Planned versus Actual Development
Expenditures, 1959-64
(millions of Iraqi dinars)

Sector	1 Planned	2 Actual	3 2 to 1 (percent)
Agriculture	131.3	42.3	32.2
Industry	154.5	50.2	32.5
Transport and communication	199.8	76.5	38.3
Buildings and services	270.4	139.4	51.5
Total	756.0	308.4	40.8

Source: Economic Plans in Iraq, pp. 46, 49.

Comprehensive National Planning

The era of comprehensive planning began in 1965 when a new economic plan for fiscal years 1965-69 was introduced.[18] In drawing up this plan, the planner was guided by three main principles, which were: (1) to raise appreciably the standard of living of the Iraqi people by accelerating the rate of economic growth, (2) to restore equilibrium to the Iraqi economy and reduce dependence on oil revenues by diversifying investment and production, and (3) to increase the level of output of the commodity sectors by ensuring higher rates of growth in both agriculture and industry. Specifically, the planner proposed the minimum annual rates of growth should be as follows: national income 8 percent; agriculture, 7.5 percent; industry, 12 percent; and the commodity sectors as a group 9.4 percent.[19]

The necessary investment for achieving these targets was estimated at ID 821 million—ID 640 million of which was to come from the public sector and the remaining ID 181 million from investment by the private sector (Table 3.3). Public expenditures were to be financed by oil revenues, estimated at ID 390 million; foreign and domestic loans, estimated at ID 95 million and ID 30 million, respectively; contributions from public enterprises and government establishments, estimated at ID 90.8 million; and miscellaneous revenues, totaling ID 34.2 million.[20]

TABLE 3.3

Iraq: The Five-Year Economic Plan, 1965-69
(millions of Iraqi dinars)

Sector	Public Sector*	Private Sector	Total	Percent of Total
Agriculture	145.0	12.0	157.0	19.1
Industry, electricity, and water	210.0	5.0	215.0	26.2
Transport and communication	99.0	20.0	119.0	14.5
Buildings, housing, and social services	122.5	141.0	263.5	32.1
Trade and services	1.0	3.0	4.0	0.5
International obligations	25.0	—	25.0	3.0
Planning and follow-up systems	2.5	—	2.5	0.3
Ministry of Defense production projects	35.0	—	35.0	4.3
Total	640.0	181.0	821.0	100.0

*Central government, municipalities, and public enterprises.
Source: Republic of Iraq, Law No. 87 of 1965.

In examining the implementation of the plan, the analysis will be confined to the performance of the central government sector, where statistical data are available and fairly accurate. At the central government level, total allocations for investment expenditures during the plan years reached ID 605.5 million. Table 3.4 compares actual to planned expenditures. Rates of implementation were once again very low in all economic sectors, despite the fact that this plan was allowed to run its full course.

Implementation shortfall notwithstanding, the Five-Year Economic Plan (1965-69) contributed substantially to the maturity of development planning in Iraq. Although the plan did not reach very far in its coverage of all aspects of social and economic life, it was the first to be drawn according to the principle of comprehensive planning.

Among the steps that were taken during the period covered by the plan was the passage, in 1966, of a law broadening the authority of the Planning Board.[21] The law authorized the board to coordinate economic, fiscal, monetary, and commercial policies in order to ensure the full implementation of the plan. In addition, the board was authorized to express its opinion regarding the ordinary budget and direct the activities of the private sector to ensure their harmony with the objectives of the plan. These were concrete steps on the road toward the creation of a truly effective planning agency.

TABLE 3.4

Iraq: Central Government Actual Allocations
and Expenditures, 1965-69
(millions of Iraqi dinars)

Sector	1 Allocations	2 Expenditures	3 2 to 1 (percent)
Agriculture	173.5	56.3	32.4
Industry	187.2	103.9	55.5
Transport and communication	110.1	61.2	55.6
Building and services	134.7	66.3	49.2
Total	605.5	287.7	47.5

Source: Economic Plans in Iraq, p. 55.

On April 1, 1970, the National Development Plan (NDP) for fiscal years 1970-74 was put into effect.[22] The NDP was the first plan to consider specifically such aspects as manpower and technical training, education, research and development, foreign trade, and prices. The planner emphasized the need for expanding investment in human resources and tried to ensure a certain degree of coordination between investment and savings policies and between consumption and development needs.[23]

Regarding specific targets, the NDP was aimed at achieving the following compound annual rates of growth: agriculture, 6.9 percent; industry, 12 percent; commodity sectors as a group, 5.8 percent; distribution sectors as a group, 8.2 percent; services sectors as a group, 6 percent; gross domestic product, 6.2 percent; net national product (national income), 7.1 percent; and per capita income, 3.6 percent.[24] Total required investment was put at

ID 1,143.7 million. The central government sector was expected to contribute ID 536.9 million; public business and local administrations were expected to contribute ID 321.8 million; and the private sector was expected to contribute ID 285 million.

The plan accorded industry, mining, and electricity first priority; second in importance was the buildings and services sector; agriculture occupied third place; and the transport and communication sector was last in terms of allocations. However, in examining the planned investment of the central government, it can be seen that first priority was accorded the agricultural sector, as it received 34.4 percent of the total; industry received 24.6 percent; buildings and services received 14 percent; and transport and communications received 11.2 percent (Table 3.5).

Thus was the original version. The rise in oil revenues following the Tehran Agreement in 1971 led to a formal amendment of the plan whereby the central government's investment target was increased to ID 952.5 million.[25] Similarly, the fourfold increase in oil prices prompted the government to increase vastly its investment expenditures during 1974, the last year in the life of the plan. Planned investment expenditures during that year amounted to ID 1,169 million.[26] These revisions raised the allocations for the five-year period from ID 536.9 million to ID 1,932 million.

Although the plan was not revised in a formal way, the 1974 Investment Program resulted in a change of emphasis, with industry emerging as the main beneficiary. While industry appeared to be receiving only 20 percent of the total over the plan period, its actual share was close to 40 percent. The reason lies in the fact that most of the additional allocations in the 1974 Investment Program under the category of "other" was earmarked for oil, petrochemical, and other major industrial projects. The increase in the industrial sector's allocation seems to have been accomplished at the expense of agriculture, whose final share declined to 19.0 percent (Table 3.6). This however, does not imply changes in the long-term development policy objectives. Rather, it is a reflection of the fact that it is more difficult to increase investment in agriculture than in industry.

During the life of the NDP (1970-74), central government expenditures on the four main sectors amounted to ID 886 million. The sectoral distribution was as follows: ID 208 million for agriculture, ID 330 million for industry, ID 177 million for transport and communication, and ID 171 million for buildings and services. Comparison between actual expenditures and actual allocations yields the following rates of implementation: overall, 70.4 percent; agriculture, 56.8 percent; industry, 84.4 percent; transport and communications, 80.8 percent; and buildings and services, 60.4 percent (Table 3.7).

TABLE 3.5

Iraq: National Development Plan (1970-74)—Distribution of Total Investment (millions of Iraqi dinars)

Sector	Central Government	Public Business and Local Administration	Private Sector	Total	Percent of Total
Agriculture	185.0	8.0	18.0	211.0	18.5
Industry, mining, and electricity	132.0	212.0	50.0	394.0	34.5
Transport and communication	60.0	54.3	35.0	149.3	13.0
Building and services	75.0	47.5	182.0	304.5	26.6
International obligations	44.0	—	—	44.0	3.8
Loans for government departments	27.3	—	—	27.3	2.4
Miscellaneous investment expenditures	13.6	—	—	13.6	1.2
Total	536.9	321.8	285.0	1,143.7	100.0

Source: Republic of Iraq, Law No. 70 for 1970.

TABLE 3.6

Iraq: National Development Plan (1970-74)—Central Government Annual Investment Programs
(millions of Iraqi dinars)

	1970	1971	1972	1973	1974	Total	Percent of Total
Agriculture	28	60	23	65	190	366	19.0
Industry	28	50	28	60	225	391	20.2
Transport and communication	15	28	16	40	120	219	11.3
Buildings and services	13	28	22	45	175	283	14.7
Other*	32	36	46	100	459*	673	34.8
Total	116	202	135	310	1,169	1,932	100.0

*Includes international obligations and loans to government agencies. For 1974 this item also includes allocations for the development of the northern region and major development projects (principally petro-chemical projects).

Source: Iraq, CSO, *Annual Abstract of Statistics 1976*.

TABLE 3.7

Iraq: Central Government Actual Allocations
and Expenditures, 1970-74
(millions of Iraqi dinars)

Source	1 Allocations	2 Expenditures	3 2 to 1 (percent)
Agriculture	366	208	56.8
Industry	391	330	84.4
Transport and communication	219	177	80.8
Buildings and services	283	171	60.4
Total	1,259	886	70.4

Source: Iraq, CSO, Annual Abstract of Statistics 1976.

EVALUATION OF DEVELOPMENT PLANNING

This section evaluates economic development planning in Iraq against a background of accepted theoretical principles. Specifically, it deals with aspects of strategy, mobilization and allocation of resources, and plan implementation.

The Strategy of Economic Development

Iraq opted from the start for a strategy of balanced growth. This is evident from the emphasis all economic programs and plans placed on the need to strike a balance between the levels of investment in the various economic sectors. The choice did not stem from a commitment to the doctrine of balanced growth.[27] Rather, it was a common-sense decision consistent with the adequate financial resources of the country and the desire to diversify its economy and lessen its dependence on oil. Be that as it may, the strategy of balanced growth was not always adhered to, and, today, the country is more dependent on oil revenues than in 1950 when the DB was established.

The original program approved by the DB contained no allocations for industry.[28] In drawing subsequent programs, the DB was influenced heavily by foreign consultants whose advice was predominantly against what some of them termed "premature" industrialization.[29] Others suggested that the reluctance of the DB to accord industry the attention it deserved was a deliberate policy action so as not to disturb the existing relationships among major social classes.[30]

Aware of the criticism leveled at the DB, the revolutionary regime that came to power in 1958 accorded industry a position of top priority in the DEP. Buildings and services came out second; the transport and communication sector was a very close third, whereas agriculture was relegated to last place. Except for a slight modification, namely, the elevation of agriculture to second place, the pattern of allocating public sector investment funds was continued under the Five-Year Plan (1965-69). Under the NDP (1970-74) the order of priority at first went as follows: agriculture, industry, buildings and services, and transport and communications. However, subsequent revisions tilted the balance once again in favor of industry.

Examination of actual central government investment expenditures during the 1951-74 period confirms the bias in favor of industry and buildings and social services to the detriment of agriculture and transport and communication. It is shown in Table 3.8 that, over this 25-year period, industry received 30 percent; buildings and services, 26 percent; and agriculture and transport and communications sectors, 22 percent each. This indicates that the distribution of expenditures was unbalanced during the three phases through which economic development planning in Iraq has passed. Consequently, the economy's dependence on oil revenues has become greater over time. It is not difficult to explain the causes that underlie this unplanned development.

TABLE 3.8

Iraq: Sectoral Distribution of the Central Government
Actual Investment Expenditures, 1951-74
(millions of Iraqi dinars)

	Amount	Percentage of Total
Agriculture	388	22
Industry	519	30
Transport and communications	382	22
Buildings and services	458	26
Total	1,747	100

Sources: Compiled from data presented in Tables 3.1, 3.2, 3.4, and 3.7.

First, traditionally, Iraq has been an agricultural country with wheat, barley, and dates as its major exports. Several causes coincided, however, to retard the growth of the agricultural sector during the last two decades. Chief among these causes, undoubtedly, was the decision of the planner during the 1950s to concentrate on dams and flood-control schemes to the neglect of drainage canals, land reclamation, development of animal wealth, extension services, and other activities that directly contribute to increasing agricultural output. Although this weakness was recognized by subsequent planners, sufficient remedies were not provided because of the preoccupation with industrialization.[31]

Stagnating agriculture, growing population, and the need to provide food through market channels to an evergrowing number of city dwellers with rising per capita income transformed Iraq from an exporter into an importer of foodstuffs. This development led to a greater dependence on oil because "the only substitute for higher agricultural production is increasing dependence on food imports which signifies greater reliance on oil revenues."[32]

Second, another contributing factor to the continuing dependency on oil has been the failure to develop export-oriented industries. Iraq, like many developing countries, pursued import substitution.[33] Import substitution certainly has been effective in Iraq, and, to this extent, no criticism is warranted. However, the planning machinery neglected export-oriented industries even in areas where the country was, and still is, admirably endowed. These include the full development of natural resources, particularly oil and sulfur. "Until the Iraqis take in hand the full development of these two mineral resources and promote the multitude of chemical industries associated with them, industrialization will remain an anemic process."[34]

The tendency to concentrate on import substitution was encouraged from the very beginning by foreign advisers. The World Bank Mission advised that "further development of the oil industry can safely be left to the internationally-owned companies now operating in Iraq."[35] Arthur D. Little, Inc., which was engaged for the specific purpose of surveying Iraq's industrial potential, called attention to the danger of the small size of the domestic market and recommended a go-slow policy in the development of petrochemical industries.[36]

The failure to develop oil and sulfur at an early stage proved to be costly. While hesitating to establish petrochemical projects, the country, inadvertently, delayed the process of its own industrialization. This is so because it is an established fact that this industry is well suited for inducing industrial expansion. "Owing to its many sided forward linkages, this industry, which proved to be among the most dynamic factors of postwar industrial growth in developed countries, can pave the way for the development of a broad range of manufacturing industries."[37]

Mobilization and Allocation of Resources

Oscar Lange points out that:

The problems which planning faces can be divided into two categories. One is the mobilisation of resources for purposes of productive investment, the other is the direction of the investment into proper channels. These are the essential problems implied in planning.[38]

Given these considerations, it follows that the planner must explore all possible sources of finance, base public investment decisions on the well-recognized criterion of social marginal cost versus social marginal revenue,[39] and devise measures, such as the use of the price mechanism as Lange suggests, to direct private investment toward those activities for which it is best suited.

Regarding mobilization, it can be said that, although Iraq has a relatively long history in development planning, the record shows that it has not been efficient in mobilizing all available resources. Until very recently, the planner exerted no effort to encourage private domestic investment. Indeed, one of the criticisms leveled against the DB was the failure to take interest in the private sector.[40] Like its predecessor, the revolutionary regime that succeeded the DB assigned no specific role to the private sector.

The Five-Year Plan (1965-69) was the first to recognize explicitly the role of the private sector by calling upon it to provide ID 181 million, or 22 percent of total anticipated investment. This was a step in the right direction, despite the fact that there was no clear indication of how this target was to be achieved. It was not until the NDP (1970-74) was announced that the government explained that the private sector had a role to play in the areas of consumer goods and services and that the public sector would not be allowed to compete in these areas.

The failure of the planner to mobilize domestic private capital was outranked by his failure to attract and mobilize foreign capital.[41] The DB never entertained the idea of attracting foreign investment or loans. In an obvious change of policy, the revolutionary regime sought foreign loans to finance projects included in

The DEP.* The planner in charge of the Five-Year Plan (1965-69) decided that "a proportion not exceeding 12 per cent of the grand total of both public and private sectors investments will be financed by foreign loans. . . . Therefore it was decided to limit financing by foreign loans to ID 95 million."[42] While stating his determination to push economic development forward "beyond the limits imposed by the State's own resources . . . through loans from friendly countries,"[43] the drafter of the NDP (1970-74) limited financing by foreign loans to ID 60 million.

If success eluded Iraqi planners in the area of mobilizing resources, it also eluded them in the area of directing the resources they had under their disposal into proper channels. A survey of all programs and plans approved by planning agencies in Iraq shows that sophisticated techniques were not employed to rationalize investment. The programs of the DB were no more than collections of investment projects, sometimes chosen on the basis of engineering feasibility rather than economic efficiency.[44] The PEP, which was adopted as an interim measure, contained most of the unfinished projects of the DB era.

The DEP was the first in Iraq's history to have a specific target, namely, the doubling of national income in a ten-year period. Taking into account the required rate of growth and an arbitrary overall capital-output ratio of 4, the planner estimated that a grand investment of ID 566.3 million was required to achieve the desired target.[45]

More sophistication was exhibited in the preparation of the Five-Year Plan (1965-69); global as well as sectoral growth targets were specified. There are, however, two observations to be made concerning these targets. First, some of them were unrealistic in view of past performance. For example, although the rate of growth in the agricultural sector was no more than 2.7 percent between 1953 and 1963, the planner expected agricultural output to grow at a compound annual rate of 7.5 percent between 1965 and 1969.

Second, the targets convey a sense of precision that was, in fact, missing when the plan was prepared. The projected increases in sectoral value added hinged upon the realization of planned sectoral investment as well as upon the validity of the assumptions and criteria according to which investment allocations were made. The

---

*The proceeds of a Soviet loan (ID 65.2 million) and a Czech loan (ID 12.0 million) were among the sources of finance listed in this plan.

plan estimated that a gross investment of ID 813 million was needed for achieving production targets. This estimate was arrived at through the utilization of the Harrod-Domar growth model, under the assumption that the overall capital-output ratio would be 3.6 during the period of the plan.[46] In allocating total investment among the principal sectors of the economy, the planner made use of the relation between the sectoral capital-output ratios and the desired sectoral value added. No hard evidence was advanced in support of the assumptions about the values of capital-output ratios.

The NDP (1970-74) considered not only the economic activities of the public sector but also those of the private sector. The plan aimed at coordinating investment activities, consumption, production, employment, foreign trade, and other economic activities in a manner that would ensure the achievement of certain economic and social targets. Another positive feature of the plan was its attempt at defining fiscal, monetary, and commercial policies in preparation for their use in controlling government consumption expenditures, encouraging the private sector to play a more active role, achieving a balance between resources and requirements of the plan, and maintaining stable general price levels.[47]

The principal target of the plan was to increase per capita income by an annual compound rate of growth of 3.6 percent. Given the rate of population growth, the achievement of this target hinged on the ability of the GNP to grow at an annual compound rate of 7.1 percent, and this, in turn, hinged on the realization of the projected sectoral rates of growth. In order to secure the material needs of the projects included in the plan, the planner reckoned a gradual change in the composition of imports by increasing the relative importance of investment and intermediate goods from 65 percent in 1969 to 75.2 percent in 1974.[48]

The targets of the plan, though more realistic, were arrived at in the same manner as in the previous plan. Output targets were set exogenously, and, following the example of the Harrod-Domar model, total investment and its allocation among the main sectors were determined. As in the previous plan, the overall and sectoral capital-output ratios were no more than adjustable factors. Nowhere in this plan could one find convincing reasoning justifying the projected targets. Nor was there any empirical evidence to substantiate the presumed capital-output ratios.

Although it is true, as W. Arthur Lewis and others once stated, that the use of advanced techniques is not necessarily useful, especially in the case of small and developing countries,[49] it is certainly true that their use would not be detrimental to the development effort.

Plan Implementation

Table 3.9 summarizes the rates of implementation achieved under the various economic development programs and plans. These ratios reflect the lag between performance and expectation and demonstrate once again Iraq's disappointing record in the field of development planning.

Some of the difficulties underlying the low rates of completion included such factors as the lack of a viable administrative system, the shortage of skilled manpower, and the poor coordination between the planning apparatus and the various government agencies.[50] In addition, there were other major factors responsible for this poor performance. The fact that planning was not comprehensive made the appearance of shortages and bottlenecks not only possible but likely.

Another problem certainly has been the incompetence of planning organs in the various ministries and the inability of the central planning machinery to provide them with proper direction and assistance. Lacking professional planning staff, many ministries submitted, for inclusion in the development plans, projects that received little study and evaluation. The government tried to deal with this weakness by passing a law providing for the establishment, within each ministry, of a department of planning and follow-up.[51] These units were intended to assist the Planning Board in the preparation of the development plan and the annual investment programs and follow up their implementation. However, "despite the passage of a long time since this law was approved, these departments were not established and no positive steps were taken to implement the law."[52]

Lack of control over ordinary expenditures was particularly damaging to the development effort. Instead of viewing the ordinary budget as an instrument for achieving the targets of the development plan, the latter was made an adjustable instrument to accommodate the needs of current government expenditures. During the period covered by the Five-Year Plan (1965-69), for example, a total of ID 77.5 million was transferred from the development budget to the ordinary budget.[53] This amount was equal to 27 percent of the total of actual expenditures under the plan. Clearly, this was contrary to sound practices in the field of development planning.

SUMMARY AND CONCLUSION

During the 1951-74 period, eight economic programs and plans were adopted in Iraq. The average life-span of a typical plan

TABLE 3.9

Iraq: Indexes of Plan Implementation, 1951–74
(percent)

	Agriculture	Industry	Transport and Communication	Buildings and Services	Overall
Revised First General Program	49.6	18.4	55.5	68.0	49.1
Revised Second General Program	44.4	56.5	46.9	76.9	54.2
Provisional Economic Plan	51.1	36.4	25.1	34.6	34.5
Detailed Economic Plan	22.7	31.5	47.9	73.6	45.3
Five-Year Plan (1965–69)	32.4	55.5	55.6	49.2	47.4
National Development Plan (1970–74)	56.8	84.4	80.8	60.4	70.4

Sources: Economic Plans in Iraq, pp. 8, 19, 46, 59; and Tables 3.4 and 3.7.

was about three years. With the exception of the last two plans, every other plan or program was either suspended or canceled. Furthermore, actual performance has consistently been short of expectation. The overall rate of implementation during the 1951-74 period was 58 percent. Sectoral rates were as follows: 46 percent in agriculture, 64 percent in industry, 57 percent in transport and communications, and 59 percent in buildings and services.

Low implementation rates were not caused by lack of funds or ready-to-go projects. Rather, they were a manifestation of the low absorptive capacity of the Iraqi economy. In particular, shortages of skilled labor and professionals, poor planning, poor administration, and political instability were detrimental to the development effort. These and other absorptive capacity constraints will be examined further in the next four chapters.

NOTES

1. Government of Iraq, Law No. 23 of 1950. It should be mentioned that starting in "1934 most of the oil royalties were utilized to finance a supplementary budget . . . the Capital Works Budget." See Ribhi Abu el-Haj, "Capital Formation in Iraq, 1922-1957," Economic Development and Cultural Change 9 (1961): 605.
2. Albert Y. Badre, "Economic Development in Iraq," in Economic Development and Population Growth in the Middle East, ed. Charles A. Cooper and Sidney S. Alexander (New York: American Elsevier, 1972), p. 286.
3. Government of Iraq, Law No. 35 of 1951.
4. Government of Iraq, Law No. 25 of 1952.
5. Government of Iraq, Law No. 35 of 1953.
6. Government of Iraq, Law No. 43 of 1955.
7. Lord (James Arthur) Salter, The Development of Iraq: A Plan of Action (Baghdad: Iraq Development Board, 1955).
8. Arthur D. Little, Inc., A Plan for Industrial Development in Iraq (Cambridge, Mass.: Arthur D. Little, 1956).
9. Government of Iraq, Law No. 54 of 1956.
10. Ibid.
11. Republic of Iraq, Law No. 74 of 1959.
12. Republic of Iraq, Law No. 181 of 1959.
13. Ibid.
14. Republic of Iraq, Law No. 70 of 1961.
15. Badre, "Economic Development of Iraq," p. 288.
16. Republic of Iraq, Law No. 74 of 1959.
17. Republic of Iraq, Law No. 24 of 1964.
18. Republic of Iraq, Law No. 87 of 1965.

19. Ibid., The Explanatory Memorandum, pp. 72-73.
20. Ibid., table 10 (a), p. 84 and table 11, p. 88.
21. Republic of Iraq, Law No. 18 of 1966.
22. Republic of Iraq, Law No. 70 of 1970.
23. Ibid., p. 133.
24. Ibid., pp. 146-47.
25. Republic of Iraq, Law No. 158 of 1971.
26. Iraq, Ministry of Planning, The Investment Program for the Year 1974-75 (Baghdad, Iraq, 1975). (Arabic.)
27. Balanced growth is based on the premise that developing countries suffer from low productivity and per capita income. This fact rendered the typical size of their domestic markets too small to permit advantageous utilization of large-scale production. In order to break this vicious circle they need to apportion their investment funds among the various sectors—that is, follow a strategy of balanced growth so as to insure an overall enlargement of the market.

The doctrine of balanced growth was challenged on the grounds that developing countries, suffering as they are from lack of funds and entrepreneurial talent, are in no position to establish several industries simultaneously. Given their circumstances, it was argued that it would be preferable for these countries to pursue a strategy of unbalanced growth—that is, to create deliberately and maintain tensions, disproportions, and disequilibria by concentrating investment in certain growing sectors.

On balanced growth see Ragnar Nurkse, Problems of Capital Formation in Underdeveloped Countries (Oxford: Basil Blackwell, 1953). On unbalanced growth see Albert O. Hirschman, The Strategy of Economic Development (New Haven, Conn.: Yale University Press, 1958). For an appraisal of these two doctrines see Paul Streeten, "Balanced versus Unbalanced Growth," The Economic Weekly, April 20, 1963, pp. 669-71; and Wilfred A. Ndongko, "'Balanced' versus 'Unbalanced' Growth as International Development Strategies," Mondes en Developpement, no. 13 (1976), pp. 904-10.

28. The inaugural program of the DB consisted of flood control projects, which had already received much study and for which plans were almost completed before the creation of the board. See Abbas Alnasrawi, Financing Economic Development in Iraq (New York: Praeger, 1967), p. 67.

29. See IBRD, The Economic Development of Iraq (Baltimore: Johns Hopkins University Press, 1952); and Arthur D. Little, Inc., Plan for Industrial Development in Iraq; and Salter, Development of Iraq: A Plan of Action. For a critical examination of the views expressed in the references mentioned above, see Taher H. Kanaan,

Input-Output and Social Accounts of Iraq 1960-1963 (Baghdad, Iraq: Ministry of Planning, 1965), pp. II.8-II.15; and Alnasrawi, Financing Economic Development in Iraq, pp. 75-83.

30. Jawad Hashim, Hussein Omar, and Ali al-Munoufy, Evaluation of Economic Growth in Iraq: 1950-1970, vol. 1, The Planning Experience (Baghdad, Iraq: Ministry of Planning, 1970), pp. 46-48 (hereafter cited as The Planning Experience).

31. Industrialization causes a diversion of capital from agriculture to industry, which leads to a maldistribution of income in favor of urban areas and encourages migration to cities. It also creates pressure on prices and the balance of payments and expands the demand for social services. Iraq has been able to lessen the impact of these problems by following a liberal import policy made possible by the continuous rise in oil revenues. For a brief but concise survey of the effects of industrialization policies see Ian Little, Tibor Scitovsky, and Maurice Scott, Industry and Trade in Some Developing Countries (New York: Oxford University Press, 1970), pp. 6-11.

32. Marwan Iskandar, "Economic Development Plans in Oil Exporting Countries and Their Implications for Oil Production Targets," in Continuity and Change in the World Oil Industry, ed. Zuhayr M. Mikdashi, Sherrill Cleland, and Ian Seymour (Beirut, Lebanon: Middle East Research and Publishing Center, 1970), p. 46.

33. The enthusiasm for import substitution stems from the desire to emulate the experience of industrialized countries. In addition, there is the argument advanced by Raul Prebisch, who contends that a "peripheral" country's demand for industrial imports increases much more rapidly than does the foreign demand for its exports. Accordingly, the country must produce the industrial goods that it cannot import, owing to the slow growth of its exports. Other arguments favoring import substitution include the need to expand employment opportunities outside the agricultural sector and the desire to attract foreign capital. For a summary of these views and a criticism thereof, see Gerald E. Meier, Leading Issues in Economic Development, 2d ed. (New York: Oxford University Press, 1970), pp. 402-9, and 484-92.

34. Badre, "Economic Development in Iraq," p. 298.

35. IBRD, Economic Development of Iraq, p. 280.

36. Kathleen M. Langley, The Industrialization of Iraq (Cambridge, Mass.: Harvard University Press, 1962), p. 219.

37. United Nations, Economic and Social Office in Beirut, Studies on Selected Development Problems in Various Countries in the Middle East (New York: United Nations, 1967), p. 21. Quoted in Badre, "Economic Development in Iraq," p. 299.

38. Oscar Lange, "Economic Development, Planning and International Cooperation" (Cairo: Central Bank of Egypt, 1961); reprinted in Meier, Leading Issues, p. 696.

39. The use of the social marginal productivity criterion was first advocated by A. E. Kahn; H. B. Chenery applied it to a number of empirical situations in several countries. See A. E. Kahn, "Investment Criteria in Development Programmes," Quarterly Journal of Economics 65 (1951): 38-61; and H. B. Chenery, "The Application of Investment Criteria," Quarterly Journal of Economics 67 (1953): 76-96.

Economists differ as to which criterion for allocation investment resources is best. For a survey see United Nations, ECAFE, "Criteria for Allocating Investment Resources among Various Fields of Development in Underdeveloped Countries," Economic Bulletin for Asia and the Far East, June 1961, pp. 30-33. Reprinted in Meier, Leading Issues, pp. 340-44.

40. Langley, Industrialization of Iraq, pp. 235-36.

41. It is clear that absorptive capacity constraints, as evidenced by low implementation rates and persistent surplus of actual revenues over actual development expenditures, convinced Iraqi planners that there was no need to mobilize foreign sources of finance. It is equally clear that those planners were missing the point. Foreign finance brings with it technology and expertise, two necessary ingredients that Iraq needed most in order to expand its absorptive capacity. In order to rectify this shortcoming, Iraq concluded several agreements involving the exchange of oil and oil products for goods and services. The most outstanding of these were the two low-interest credit agreements, worth $1 billion each, that were concluded with Japan in January 1974 and January 1977. Their proceeds will be utilized in the implementation of a number of oil development and industrial projects on a turnkey basis. See Middle East Economic Survey, January 13, 1977, p. 9.

42. Republic of Iraq, Law No. 87 of 1965: The Explanatory Memorandum, p. 87.

43. Republic of Iraq, Law No. 70 of 1970: Explanatory Note, p. 136.

44. Hashim, Omar, and Ali al-Munoufy, The Planning Experience, p. 47.

45. Republic of Iraq, Law No. 70 of 1961, pp. 48, 67.

46. Republic of Iraq, Law No. 87 of 1965: The Explanatory Memorandum, p. 77.

47. Republic of Iraq, Law No. 70 of 1970: Explanatory Note, pp. 132-37.

48. Ibid., p. 164. The ratio of the total value of imported principal capital goods to the total value of imports was as follows:

58.7 percent in 1972, 57.1 percent in 1973, 56.3 percent in 1974, and 55.0 percent during January-September 1975. See CBI, Bulletin, no. 1, January-March 1976, pp. 61-63.

49. W. Arthur Lewis, "On Assessing a Development Plan," Economic Bulletin of the Economic Society of Ghana, May-June 1959, pp. 2-16. Reprinted in Meier, Leading Issues, pp. 716-22.

50. These and other difficulties to be discussed below are common to most developing countries. See Albert Waterston, Development Planning: Lessons of Experience (Baltimore: Johns Hopkins University Press, 1965), chap. 9, particularly pp. 314-32.

51. Republic of Iraq, Law No. 27 of 1965.

52. Hashim, Omar, and Ali al-Munoufy, The Planning Experience, p. 83.

53. Jawad Hashim, Hussein Omar, and Ali al-Munoufy, Evaluation of Economic Growth in Iraq: 1950-1970, vol. 2, The Evolution of the Commodities Sectors (Baghdad, Iraq: Ministry of Planning, 1970), p. 510 (hereafter cited as Commodities Sectors).

4

PERFORMANCE OF
THE IRAQI ECONOMY
UNDER DEVELOPMENT PLANNING

The overriding objective of development policy in Iraq has been, and still is, the stimulation of the nonoil sectors in order to reduce the country's heavy dependence on oil revenues. This chapter examines the performance of the Iraqi economy during the last two decades, that is, since the introduction of formal development planning. The analysis will proceed from the aggregate to the sectoral aspects with special emphasis on the commodity producing sectors.

OVERALL PERFORMANCE

Table D.3 presents Iraq's gross domestic product by economic sector at constant 1969 prices for the 1953-73 period. Data in this table were used to estimate aggregate and sectoral rates of growth over the 1954-73 and 1964-73 periods. Ordinary least squares method of estimation was employed in the performance of this exercise.* Estimated rates of growth are shown in Table 4.1; Table D.6 contains more detailed information on the results of the regression.

Examination of the data presented in Table D.3 and Table 4.1 leads to the following generalizations:

---

*The procedure employed in this instance as well as in estimating aggregate and sectoral rates of growth of investment in Iraq is explained in Appendix B.

TABLE 4.1

Aggregate and Sectoral Real Growth Rates of the Iraqi Economy
(in percent)

	1954-73	1964-73
Overall GDP	6.03	4.72
Nonoil GDP	6.76	5.21
Nonoil, nonagricultural GDP	7.07	5.97
GDP in the agricultural sector	4.54	2.81
GDP in the mining and quarrying sector	4.68	3.80
GDP in the industrial sectors	6.68	7.21
GDP in the distribution sectors	6.86	3.96
GDP in the services sectors	8.96	6.54

Source: Derived from Table D.3.

1. Gross domestic product (expressed in terms of constant 1969 prices) rose from ID 406.6 million in 1953 to ID 1,360 million in 1973, implying an average rate of growth of about 6 percent a year. Below average growth (around 4.5 percent) was experienced by the two leading sectors, oil and agriculture; on the other hand, above average rates of growth were obtained by the industrial, distribution, and services sectors.

2. The overall rate of growth was lower in the 1964-73 period than that achieved over the 20-year 1954-73 period. This was due largely to the sluggish growth in the two leading sectors, oil and agriculture. The lackluster performance of the oil sector was due to the stagnation of oil production (discussed in Chapter 2) and to the decline in the real value of oil. Specifically, oil prices declined in nominal terms in the 1950s and were essentially unchanged in nominal terms throughout the 1960s. The terms of trade had definitely shifted against Iraq and other major oil producers during this period.

The slow growth in agricultural output largely reflected inadequate planning, unfavorable weather conditions, political instability, and difficulties that delayed the implementation of agrarian reform, thus adversely affecting agricultural output.

3. Owing to the poor performance of agriculture, the contribution of the commodity producing sectors other than oil to GDP declined from 32.5 percent in 1953 to 29.5 percent in 1973. In contrast, the relative share of oil-dependent sectors had increased. Public administration and defense, for example, accounted for 5.7 percent of

GDP in 1953, 10.0 percent in 1963, and 11.3 percent in 1973. Similarly, the contribution of the distribution sectors as a group rose from 18.4 percent in 1953 to 22.3 percent in 1973.

4. The decline in the share of the commodity producing sectors (other than oil) and the rise in the share of the services and distribution sectors meant heavier reliance on crude oil exports. The serious imbalance in the growth of the economy is demonstrated by the fact that exports other than crude oil experienced little or no growth, whereas imports increased several-fold during the 1950-73 period. In brief, Iraq has been transformed from an exporter of agricultural products into a food importer. Exports other than oil, which financed about 68 percent of the country's imports in 1950, financed only about 12 percent of total imports in 1973.

5. As pointed out in the preceding chapter, these results run contrary to the stated objective of development planning in Iraq, namely, the achievement of balanced growth. Failure to adhere to this policy did not only lead to deeper reliance on oil revenues but also paved the way for the emergence of certain bottlenecks such as the transportation crisis of 1974 and 1975 and the persistent shortage of skilled manpower.

SECTORAL PERFORMANCE

Agriculture

Next to oil, agriculture is the most important sector in the Iraqi economy, contributing in 1973 about 13.3 percent of GDP and providing jobs for about 56 percent of employed labor in the country.

Cultivable land in Iraq is estimated at 12 million hectares or a little more than one-fourth of the total area of the country.[1] However, owing to the widespread practice of the fallow system, only about a fourth of the cultivable land is actually cropped in any given year.[2] Clearly, a major objective of agricultural development must be the expansion of cropped area. This can be accomplished by discarding the fallow system and/or by reclaiming more land.

The bulk of cultivable land is held in small farm units, following passage of the agrarian reform program in 1958; prior to that, feudalism was the norm. Under the feudal system, the distribution of land was quite skewed, with a few landlords owning the bulk of it.

As previously indicated, agriculture failed to match the rate of growth attained by the national economy throughout the last two decades. Several factors have contributed to the poor performance of this sector, among which are the following.

Soil Salinity

Because of poor irrigation practices and past neglect and mismanagement, a large part of irrigated land in Iraq today is affected by salinity and waterlogging. "The problem that normally ensues after completing the irrigation networks and the beginning of the utilization of agricultural lands is the appearance of salinity on the surface of the soil."[3]

So far, investment in main drainage canals seems to be insufficient to complement the established irrigation network. The problem was compounded by the fact that the government did very little in the way of on-farm drainage and soil improvement, assuming that such complementary investment would be undertaken by individual farmers. As events have shown, the assumption was a false one.

Thus, land utilization has been minimized and productivity per unit of area has declined to a level that is considered to be one of the lowest in the world. Taking wheat as an example, the average yield of one hectare during the 1962-68 period was 530 kilograms in Iraq compared with 1,380 kilograms in Australia, 1,760 kilograms in the United States, 2,090 kilograms in Yugoslavia, and 2,640 kilograms in Egypt.[4]

The adverse effect of salinity and waterlogging is painfully felt in yet another way. In recent years the government has been forced to reduce gradually the area reserved for rice paddies in an effort to ameliorate the salinity problem.[5]

Another problem closely connected with soil salinity and waterlogging is the practice of the fallow system. Land is left temporarily fallow in the belief that this exercise will restore its fertility. Scientific research, however, has "proved beyond a doubt that the practice of fallow neither improves land fertility nor eases the problem of water-logging, and thus, soil salinity would increase whether fallow is practiced or not."[6]

Clearly, any effort to increase agricultural production and productivity must begin with the search for ways and means of overcoming the danger of salinity and waterlogging. The first step to be taken is the completion of the main drainage system and the establishment of on-farm drainage facilities.

Land Reform

Though it was a just and necessary measure, the agrarian reform that was enacted in 1958 has been responsible, at least partially, for the decline in agricultural production:

> The reform reduced production, first, through the extreme uncertainty caused by delay in distribution; on

requisitioned land owners did not cultivate more than they expected to retain, while the cultivators did not know, and for the most part still do not know, what land would be allocated to them. Second, it reduced production in the irrigation zone through the failure to replace land owners' functions in pump maintenance.[7]

Land distribution fell considerably behind schedule. This was occasioned by the need to complete surveys, improve the sequestrated land, select suitable farmers, and establish cooperatives and a host of other requirements.[8] Farmers, having no security of tenure, either ignored good farming practices or left the farms altogether and moved to the cities. These problems still persist even today, a fact that prompted the government in January 1974 to pass a new law designed to amend the land distribution procedures.[9]

The new law permits land sequestrated under the agrarian reform laws of 1958 and 1970 to be leased for cultivation by individual farmers and cooperatives, thus making it possible to bypass the formalities of the agrarian reform laws. With this law the government hopes to solve the problem of poor farming practices and reduce migration to urban areas by inducing farmers who had cultivated the land to remain there. The government hopes that with the provisions of the new law cultivated areas as well as agricultural output will be expanded.

Other Obstacles

Lack of trained personnel is a basic problem in agriculture. The agrarian reform law of 1958 left void the role previously played by the landlord. To remedy the situation the government attempted to form farm cooperatives but with limited success. For example, the government called for the creation of 2,000 cooperatives between 1961 and 1965, but by the end of the period less than 300 were in operation.[10]

Cooperatives must be run by supervisors who are trained in agriculture, cooperative management, bookkeeping, credit, marketing, and so forth. At the end of 1963 the Ministry of Agrarian Reform established, with the help of FAO, an institute for cooperative training.[11] However, training takes time. Thus, the number of trained supervisors was only 299 by 1966.[12]

Aside from the shortage of trained personnel, the agricultural sector in Iraq suffers from general backwardness. The use of modern farm machinery, application of fertilizers, and use of improved seed varieties have been limited. Furthermore, the effort to modernize agriculture has been constrained by the lack of adequate extension service due to the scarcity of trained personnel and the lack of adequate credit facilities, because the financial resources of the Agricultural Bank have been relatively modest.

Inadequate investment is a key factor that has adversely affected agricultural production in Iraq. In fact, inadequate investment can be viewed as being closely connected with the factors mentioned above (for example, salinity, low productivity) that have limited agricultural output. The problem is a matter of implementation. Of the ID 849 million allocated by the government for agriculture over the 1951-74 period, less than half was actually spent.

Several factors coincided to slow down implementation. The practice of dividing large contracts into small ones to accommodate the limited capabilities of local contractors did not only delay implementation but resulted in the multiplication of legal and administrative procedures, which overtaxed the limited resources of concerned governmental agencies.[13] Furthermore, governmental agencies charged with the task of overseeing the implementation of investment projects were ill-equipped to handle their assignments because of the shortage of engineers, managers, and other technical staff. Delays were caused, in many instances, by a cumbersome routine faithfully adhered to by officials, especially in the Ministry of Finance. Finally, implementation of projects was often hindered because of the lack of coordination between the Ministries of Agriculture and Agrarian Reform and other ministries.

Industry

All large industrial enterprises in Iraq are government-owned, and many firms in the private sector have sizable government participation. Government interest in industry has always been paramount, and government control over industry has gradually increased. As far back as a half a century ago, the government sought to encourage and protect domestic industry from foreign competition by a combination of tax-relief measures, high tariffs, import quotas, prohibitions, and administrative restrictions on imports. Law No. 14 of 1929 for the encouragement of industries exempted profits of industrial establishments from income tax for 10 years, their property from real estate and property tax for 15 years, and their exports from export tax for an unlimited period.[14]

Following World War II, the government began to participate in industry, through the Industrial Bank, which started operations in 1946.[15] In the 1950s the government started to build directly plants through the Development Board, which was established in 1950. Following the 1958 revolution the government embarked on a major industrial program. Finally, in July 1964 the government nationalized all banks, insurance companies, and major private, industrial, and commercial establishments.[16]

To manage public enterprises (nationalized or otherwise) the government created the Economic Organization and the State Organization for Banks.[17] Less than a year later, in 1965, the Economic Organization was replaced by three public institutions: the State Organization for Industry, the State Organization for Trade, and the State Organization for Insurance.[18] Finally, in 1970 the State Organization for Industry was replaced by five state industrial organizations.[19] Each one of these organizations is in charge of a group of related industries—for example, the State Organization for Food Industries has nine subsidiaries whose products range from vegetable oils to fertilizers.[20]

Major manufacturing in Iraq is carried out by the subsidiaries of these state organizations, set up as self-accounting and semi-autonomous entities under the Ministry of Industry. The public sector concentrates on capital-intensive industrial projects with long gestation periods, leaving small-scale industries in the areas of consumer goods and services to the private sector.

Although some Iraqi manufactures such as cement and vegetable oils have entered foreign markets, the emphasis has been on import substitution. The major industries in Iraq are foodstuffs and beverages, textile and clothing, construction materials, and chemical and petroleum refining.

The industrial sector, comprising manufacturing, construction, and electricity and water, accounted in 1973 for approximately 15.5 percent of GDP and employed a total of 257,000 workers or about 9.3 percent of all employed persons in the country (Table 2.4). It is worth noting that the contribution of the industrial sector has shown an increasing trend, rising from a little less than 10 percent in 1953 to 15 percent in 1973. The bulk of the increase occurred in the manufacturing subsector whose share in GDP rose from 6.1 percent in 1953 to 10.6 percent in 1973.

The industrial sector experienced a comparatively high rate of growth during the 1954-73 period. The rate accelerated during the 1960s, and beginning in 1972 this sector has exhibited a vigorous rate of growth. Yet, considering the fact that it has enjoyed a position of top priority over the past 15 years, its performance must be judged inadequate. During the 1951-74 period, the industrial sector received ID 804 million, or about 26 percent of central government allocations. However, only ID 519 million of planned investment was actually realized.

Delays in project execution are attributable to organizational as well as to procedural problems. One of the organizational problems has been the weakness of government agencies in charge of implementing industrial projects. Most of these agencies suffered from shortages of technical staff, failed to properly supervise the progress of the work, and failed to adhere to time-phasing schedules in the implementation of integrated projects.[21] Another organizational problem stems from the failure to separate the planning function from the implementation function. Rather than entrusting it to the ministries and departments concerned, the responsibility of project execution has been shared by the planning machinery and the various ministries. In addition, there has been a lack of coordination between the executing agencies and the beneficiary agencies that will be charged with the responsibility of managing completed projects. The delivery of many projects to beneficiary agencies was delayed several years because of the lack of such coordination.[22]

Procedural problems were also responsible for the slow rate at which projects were implemented. The construction of large projects was entrusted to several contractors, many of whom lacked the necessary technical and financial competence. In many instances, contracts were awarded to the lowest bidder without regard to his qualifications. Another procedural problem was the absence of a clear and uniform accounting system according to which disbursements on a development project were to be made. This situation enabled the Directorate General of Accounts to withhold disbursements, especially when available funds were low. Financial pressures created by expansion in current expenditures frequently were relieved by the diversion of funds from the development budget to the ordinary budget.[23]

Low rates of implementation were not the only reason underlying slow industrial development. Industry in Iraq suffered from other problems, many of which are the typical kind experienced by developing countries. The small size of the Iraqi market has been a constraining factor, limiting many establishments to a level of production well below their economic capacity. This fact forced per unit cost of production up, thus making local products less competitive with imported substitutes.[24]

Shortages of qualified managers and skilled labor is believed to be the most important bottleneck the industrial sector, indeed, the economy as a whole, has to contend with. Several measures designed to deal with this problem are already underway. They range from fighting illiteracy among laborers to a counter "brain drain" program. The notable achievements in this regard include: the establishment of training institutes attached to the state industrial organizations and designed to meet their particular needs; the establishment of the State Organization for Industrial Design and Construction in charge of training industrial consultants and responsible for the design, development, and construction of new projects up to the point when they are handed over to one of the operating organizations; the training of Iraqi technicians, engineers, and managers abroad; the promulgation of Law No. 154 of 1974 for the purpose of encouraging Iraqi and Arab scientists in the United States and Europe to work in Iraq; the construction of modern, low-income housing projects near factories; and the provision of other social amenities.[25]

Import-substitution industrialization itself contributed to the retardation of the growth of industry. In a typical fashion, it created a new set of demands for a variety of imports such as machinery, spare parts, raw materials, and semifinished goods. Although crude oil exports always have generated large sums of foreign exchange, Iraq found it necessary to impose import restrictions. On several occasions the Ministry of Economics failed to allocate sufficient credits in its annual import programs to satisfy the needs of development projects.[26] The problem was compounded by the fact that the government, perhaps for political reasons, found it unwise to suppress the growing demand for imported consumer goods.

The impact of import-substitution industrialization on Iraqi imports is not unique. It is in conformity with the experience of other developing countries. Similar developments were observed in Argentina, Brazil, Mexico, India, Pakistan, Philippines, and Taiwan:[27]

> Not only does the policy of import substitution create demand for imported inputs and equipment; but the rise in per capita income is likely to raise the developing countries' propensity to import, all the more so, because, as we have seen, import restriction tends to shift the distribution of income in favour of the urban sector and higher income groups, whose expenditure pattern typically has the highest component of imports.[28]

The result of all of this has been the underutilization of productive capacity. Lack of spare parts, raw materials, or intermediate goods has often resulted in the shutdown of some factories and forced others to operate below capacity.[29]

Iraqi industry also suffered from lack of enthusiasm on the part of the private sector. Iraqi entrepreneurs are accustomed to employing their capital in trade rather than industry. The promising beginnings of the 1940s and the 1950s were reversed in the 1960s, following the 1958 revolution and the nationalization measures of 1964. To regain the confidence of private investors the present government drew a distinct line between the public and private sectors. In addition

74 / OIL REVENUES & ACCELERATED GROWTH

to the customary incentives the government had been providing over the years, investment in small and quick yielding projects, in the areas of consumer industries and tourism, was left to the initiative of the private sector. The public sector was not permitted to compete in these areas. Furthermore, the State Organization of Industrial Development was established in 1973, under the Ministry of Industry, for the purpose of planning the industrial development of the private and mixed sectors. The Industrial Bank was made a subsidiary of this organization in order to enable it to perform its duties most efficiently.[30]

As a result of the new industrial policies, the rate of growth of the industrial sectors rose at an accelerated rate approaching 50.6 percent in 1976 (Table 7.2). Be that as it may, real industrial take-off in Iraq will not materialize until the country succeeds in establishing a petrochemical industry for which it is admirably suited. Arthur D. Little, Inc. recommended in the 1950s the establishment of a small petrochemical complex,[31] yet major action was not taken until very recently.[32]

It is expected that the National Development Plan (1976-80) will accord particular priority to petrochemical, engineering and metallurgy, and construction materials industries. Refining capacity will be enlarged so as to permit increased production of fertilizers, chemicals, and petroleum derivatives. Greater fertilizer output will provide a much needed input for the agricultural sector, while chemicals and petroleum derivatives will provide inputs for the textile, chemical, plastics, and engineering industries.[33]

Industrial expansion depends on availability of electrical power. Over 99 percent of the electrical power is produced by the government-owned National Electricity Administration, which provides power to cities and industrial plants throughout the country with the exception of small municipalities.

The transmitted power rose from 1,383 million kilowatts in 1968,[34] to 2,738 million kilowatts in 1973,[35] indicating that the demand for electrical power has been growing at an annual compound rate of 14.7 percent. Though the growth rate is high, the existing capacity of 775 megawatts[36] has shown a surplus over the last few years. The National Electricity Administration, however, realized that the surplus could be exhausted rapidly as new industrial plants come into operation at an accelerated rate. To meet the anticipated rise in demand the administration is currently engaged in expanding the capacities of existing stations. More important, the Planning Board has approved the establishment of four hydro- and seven thermal stations throughout Iraq with a combined capacity of 5,830 megawatts. Some of these stations were completed in 1976, and others will be completed in 1980, all no later than 1987.[37]

Substantial investment also is planned in the construction subsector with particular emphasis on cement, bricks, concrete blocks, and plywood. Scarcity of building materials has developed into a real bottleneck during recent years, following the marked expansion of construction activity. The number of contracting firms in Iraq is relatively small, and their ability to implement their contracts has been limited by the emergence of shortages of various construction materials such as cement and bricks.

The government realizes that the implementation of the NDP depends to a certain extent on the solution of this bottleneck. This realization led the Planning Board to place special emphasis on the construction materials industries in the Investment Program for 1975.[38] To overcome the shortage in qualified contractors, the government established the State Company for Construction Contracts and the State Company for Consultancy and Implementation of Oil Projects. In addition, agreements were concluded with French firms to build factories for the production of prefabricated building materials.[39]

Considering the amount of funds allocated for construction of buildings of all types, it is natural for the authorities to be concerned with the quality of building materials and techniques. Building materials and techniques currently in use in Iraq are such that local resources are not efficiently exploited, and, as a consequence, there has been considerable waste and low standards of construction. Accordingly, a Building Research Center was established for the purpose of conducting research and providing services in order to effect economies and improve efficiency and quality in building construction.[40]

Transport and Communications

The need for a modern and efficient transport and communication system cannot be overemphasized, especially in the case of developing countries, where the services provided by this sector are vital for the speedy execution of economic development programs and the proper operation of newly established projects.

The inadequacy of the Iraqi transport system became painfully evident in 1974 when, owing to the sharp rise in oil revenues, the volume of imported goods experienced a sudden expansion. The transport system was poorly prepared to handle the influx of cargo. The problem was compounded by the limitation imposed by storage capacity, the failure of importers (the great majority are public enterprises) to remove their imported goods within a reasonable time from storage areas, and the rush to purchase most of the goods specified in the import program for 1974 during the early months of 1974.[41] Foreign ships not only crowded Iraqi, Syrian, and Lebanese ports but they had to wait weeks and months before they were unloaded.

No doubt, the rise in the volume of imports was a deciding factor in precipitating the transport crisis. Yet, the true cause always has been insufficient investment. Over the 1951-74 period this sector received ID 670 million, about 21 percent of total central government allocations. However, actual expenditures during this period were only ID 382 million. The neglect of the 1960s was continued in the early 1970s. Under the last plan only 11.3 percent of total central government allocations were earmarked for transport and communication (Table 3.6). Thus, hidden constraints surfaced when demand for the services of this sector reached its peak in 1974.

Among the problems that need to be dealt with are the elimination of the two-gauge system because it requires transshipment; renewal of rolling stock, which in its present condition hinders full utilization of existing railroad capacity; maintenance of paved roads, as over half are old and need immediate attention to meet the needs of growing traffic; better utilization of waterways; expansion of storage capacity; and better coordination between importers and transportation authorities.[42]

The government is painfully aware of the problems and is planning major action. Work is in progress to develop a second major port on Shatt-al-Arab and Umm Qasr to supplement facilities at Basrah; the construction of a sulfur depot and quays has been finished, while a modern container terminal is planned for the port; a new rail link between Baghdad and the Syrian railroad system is in the implementation stage; and contracts have been awarded for construction of a highway between Baghdad and the Syrian border. When completed these projects will ultimately link Latakia with Basrah and provide an economical land route between the Mediterranean and the Arabian Gulf.[43]

SUMMARY

This chapter reviewed the performance of the Iraqi economy under development planning. The review has shown that, although the economy has achieved an acceptable rate of growth over an extended period of time, the performance has been below its true potential. The failure to achieve a higher rate of growth is attributable in large measure to the stagnation of the oil and agricultural sectors.

Other sectors, especially industry, have performed relatively well. Nevertheless, true industrialization in Iraq still awaits the establishment of petrochemical and other heavy industries. Such industries are capital-intensive, but Iraq has adequate capital to finance them and still be able to meet the legitimate needs of other investment programs.

At the risk of pointing out the obvious, it is necessary to state that economic development is not achievable by funds alone. The country has to overcome present and potential bottlenecks—that is, it must be able to overcome the limits that restrict its absorptive capacity. It is clear from this chapter that these limits have manifested themselves in various forms. Certainly, there is a shortage of skilled managers and labor. Cumbersome administrative procedures, inefficient bureaucracy, and lack of coordination between the planning machinery and beneficiary agencies have slowed down implementation of investment projects and have led to underutilization of installed capacity. Finally and most important, political instability must be ranked as the most serious limit of the absorptive capacity of the Iraqi economy.

NOTES

1. Abdul Jalil el-Hadithy and Ahmad el-Dujaili, "Problems of Implementation of Agrarian Reform in Iraq," in Land Policy in the Near East, ed. Muhammad Riad el-Ghonemy (Rome, Italy: Food and Agricultural Organization of the United Nations, 1967), p. 219.
2. Ibid.
3. Ibid., p. 228.
4. Aziz Sibahi, "The Relationship between the Size of Agricultural Land and Increased Production," Ath-Thawra, October 20, 1972, p. 3.
5. The area planted with rice was 436,300 meshara in 1971, 376,200 meshara in 1972, and 255,800 meshara in 1973. See Iraq, CSO, Annual Abstract of Statistics 1973, p. 113.
6. Jawad Hashim, Hussein Omar, and Ali al-Munoufy, Evaluation of Economic Growth in Iraq, 1950-1970, vol. 1, Commodities Sectors (Baghdad, Iraq: Ministry of Planning, 1970), p. 100 (hereafter cited as Commodities Sectors).
7. Doreen Warriner, Land Reform in Principle and Practice (Oxford: Clarendon Press, 1969), p. 95.
8. el-Hadithy and el-Dujaili, "Problems of Implementation of Agrarian Reform in Iraq," pp. 226-27.
9. Republic of Iraq, Law No. 12 of 1974.
10. CSO, Annual Abstract of Statistics: 1973, p. 132.
11. Warriner, Land Reform in Principle and Practice, p. 97.

78 / OIL REVENUES & ACCELERATED GROWTH

12. Ibid.
13. Hashim, Omar, and Ali al-Munoufy, Commodities Sectors, p. 97.
14. Ibid., p. 142.
15. Kathleen M. Langley, The Industrialization of Iraq (Cambridge, Mass.: Harvard University Press, 1962), p. 137.
16. Republic of Iraq, Law No. 99 of 1974.
17. Republic of Iraq, Law No. 98 of 1964.
18. Republic of Iraq, Law No. 66 of 1965.
19. Republic of Iraq, Law No. 90 of 1970.
20. Iraq, Ministry of Industry, Annual Report for Fiscal Year 1973-74 (Baghdad, Iraq, 1975), pp. 53-121. (Arabic.)
21. Republic of Iraq, Law No. 70 of 1970: The Explanatory Note, p. 115.
22. Hashim, Omar, and Ali al-Munoufy, Commodities Sectors, pp. 512-13.
23. Ibid., pp. 513-14.
24. Ibid., pp. 500-1.
25. Iraq, Ministry of Industry, Annual Report for Fiscal Year 1973-74.
26. Hashim, Omar, and Ali al-Munoufy, Commodities Sectors, p. 519.
27. See Ian Little, Tibor Scitovsky, and Maurice Scott, Industry and Trade in Some Developing Countries (New York: Oxford University Press, 1970).
28. Ibid., p. 63.
29. Hashim, Omar, and Ali al-Munoufy, Commodities Sectors, p. 519.
30. Iraq, Ministry of Industry, Annual Report for Fiscal Year 1973-74, pp. 251-77.
31. Arthur D. Little, Inc., A Plan for Industrial Development in Iraq (Cambridge, Mass.: Arthur D. Little, 1956), p. 17.
32. It was reported recently that "Work has started on the first of Iraq's two major petrochemical plants, a $1.1 billion complex being built under a turn-key contract covering design, engineering, construction and commissioning by an American-West German joint venture." See Mideast Markets, March 28, 1977, p. 4.
33. See Iraq, Ministry of Industry, Annual Report for Fiscal Year 1973-74; and Iraq, Ministry of Planning, The Investment Program for the Year 1975 (Baghdad, Iraq, 1974).
34. Hashim, Omar, and Ali al-Munoufy, Commodities Sectors, p. 538.
35. Iraq, Ministry of Industry, Annual Report for Fiscal Year 1973-74, p. 285.
36. Ibid., p. 278.

37. Ibid., pp. 282-84.

38. The minister of planning pointed out, in his introduction of the 1975 investment program on Baghdad television on June 12, 1975, that the program laid special emphasis on the provision of construction materials. All brick, brick-substitutes, and other construction materials projects were approved, and their implementation has been started. See: Al-Jumhuriyah, June 13, 1975.

39. Ibid.

40. Ibid.

41. Al-Jumhuriyah, September 5, 1974. It is believed that the haste to buy early cost the government additional funds because importers did not shop around for the best possible buys, the prices of some commodities declined late in 1974, and some goods were damaged because they were left unprotected in open spaces while storage areas were full.

42. Ath-Thawra, June 18, 1974.

43. Iraq, Ministry of Planning, The Investment Program for the Year 1974-75, pp. 233-70.

5

ABSORPTIVE CAPACITY OF
THE IRAQI ECONOMY

THE CONCEPT OF ABSORPTIVE CAPACITY

After surveying the concept of absorptive capacity, including its measurement and limitations, the absorptive capacity of the Iraqi economy can then be defined and measured.

The notion of absorption refers to those limits that restrict the ability of an economy to absorb capital funds in an efficient manner. This being the case, it is often concluded that the observed low rates of investment in many developing countries are caused not by the dearth of capital funds but by their inability to utilize effectively these funds. Thus, one writer asserts that "the belief that a capital shortage is the effective or operating impediment to indigenous private investment is mistaken, that it is an illusion created by a large false demand for capital, and that what really exists is not an immediate shortage of capital at all, but a shortage of viable projects."[1]

Definition

Although references to absorptive capacity in economic literature date back at least to 1949,[2] the concept does not have an established technical meaning like "supply," "demand," or "the propensity to import."[3] Be that as it may, economists have tried to define absorptive capacity in one fashion or another. Branko Horvat defined it as follows:

> The potential effect of the optimum adjustment of the growth rates of factors is defined as the absorptive capacity of the economy. The easiest way to use this

concept is to conceive the economy as a giant productive capacity capable of being expanded at a certain <u>maximum rate</u>. Any additional inputs (investment) would not produce <u>additions</u> to but <u>reductions</u> of output. Or, applying (with caution) the conventional terminology, marginal efficiency of investment will be zero or negative.[4]

It should be pointed out here that the reference to marginal efficiency is meant to apply only to social capital. That is to say, it is entirely possible to build a modern and profitable factory (profit here means a positive financial rate of return on the particular investment) and still realize a negative social rate of return. This is so because "external diseconomies <u>with respect to the economy as a whole</u> will outweigh the positive contribution of the <u>additional factory</u>."[5]

The most specific definition is the one formulated by John H. Adler. He defined absorptive capacity as "that amount of investment or that rate of gross domestic investment expressed as a proportion of GNP, that can be made at an acceptable rate of return, with the supply of co-operant factors considered as given."[6]

Several comments need to be made at this juncture. It is clear from the definitions cited above that central to the concept of absorptive capacity is the idea of efficient utilization of capital. The criterion of "acceptable rate of return" implies that the rate of return that must be realized from installing an incremental unit of capital must at least equal the rate of return that could be realized by installing the same incremental unit of capital outside the country. Put differently, the absorptive capacity of the economy is that amount of invested capital that sets the rate of return exactly at the average rate available outside the country.

Since the rate of return on invested capital usually diminishes as more and more incremental units of capital are installed, it is also clear that the concept of absorptive capacity usually depends on the concept of diminishing marginal productivity of capital. In other words, the concept of absorptive capacity is derived from the law of diminishing returns. This law is valid for any two factors of production as long as it is assumed that one of them is fixed while the other is variable.

Since the analysis can be extended to include any number of factors of production, Adler referred in his definition to "co-operant factors," that is, to all factors of production other than capital. His definition of absorptive capacity regards the supply of cooperant factors as given. Accordingly, successive increments of capital are assumed to produce diminishing increments of output. In other words, the rate of return on invested capital is assumed to decline until it reaches the "acceptable rate." Additional investment must be regarded as inefficient because the rate of return will be less than the "acceptable rate."

This interpretation seems quite restrictive in view of the rigidity attached to the supply of cooperant factors. Such an assumption is only valid within the context of a closed economy. Otherwise, it is perfectly reasonable to assume that factors of production can be augmented through importation. Furthermore, the constancy assumption implies immobility of factors within the same economy. That is to say, a specific cooperant factor that is in constant supply is one that is relatively or completely immobile. Finally, the constancy assumption denies the possibility of substituting one factor of production for another.

Worthy of attention is the ambiguity of the definitions of absorptive capacity. A case in point is the concept of "acceptable rate of return." The rate of return is not specified. However, it can be interpreted to mean "social" rate of return as opposed to "financial" rate of return. This interpretation is adopted because to do otherwise would disregard the externalities associated with investment activities. More important, financial rate of returns usually disregard investment in human capital, which is basic to the process of economic development.

Sectoral Absorptive Capacity

It is possible to examine the concept of absorptive capacity from a sectoral perspective. The same principles that hold for the economy as a whole apply to individual sectors. Generally, the various sectors of the economy have different absorptive capacities, either because of less-than-perfect mobility of the same cooperant factor among sectors or because different cooperant factors are required by different sectors.

Examination of the absorptive capacities of various sectors makes it possible to bring into focus the limitations imposed by each sector on the other sectors. For example, one of the bottlenecks that most members of OPEC faced following the quadrupling of oil prices in 1974 was the inadequacy of ports and transportation facilities. In terms of the present analysis, the transport sector constituted a constraint on the absorptive capacity of other sectors.

The impact of each sector on the absorptive capacities of the other sectors is illustrated further by the fact that the high prices paid by other sectors for transport services would result in an increase in the absorptive capacity of the transport sector.

Consequently, additional investment would be undertaken to expand transport facilities.

Measurement

One method of measurement that readily suggests itself is that of determining the productivity of capital. That is to say, absorptive capacity can be measured in terms of rate of return on investment. If this approach is adopted, absorptive capacity becomes:

> a schedule relating an amount of capital to be invested to the expected rate of return. The lower the rate of return on capital which the "investor"—the economic unit making an investment decision—is willing to accept as satisfactory, the higher the absorptive capacity.[7]

Figure 1 illustrates this method of measurement. In this figure, investment (I) is measured along the horizontal axis, whereas the rate of return (r) is measured along the vertical axis. Curve DD relates investment to the rate of return in developed countries, and curve UU relates investment to the rate of return in developing countries. The slopes of these two curves differ because of the different conditions prevailing in developed and developing countries.

The two curves intersect at point B. To the left of B, the UU curve lies above the DD curve, signifying that investment up to I_1 would yield a higher rate of return in developing countries than in developed countries. This is so because:

> There is circumstantial and some direct empirical evidence that in less developed countries the return on existing stock of capital is high and that it is reasonable, therefore, to conclude that the expected rate of return on *some* additional investment also is high. The evidence is the larger share of profits, rents and interest receipts in the national product of many less developed countries and the high rates of interest charged by noninstitutional lenders.[8]

To the right of B, however, the situation is reversed because curve DD declines more gently than curve UU. As indicated earlier, the behavior of these curves is governed by the prevailing conditions in developed and developing countries, respectively. The steep slope of UU reflects a faster decline in the rate of return, owing to

84 / OIL REVENUES & ACCELERATED GROWTH

absorptive capacity constraints. In contrast, in developed countries where absorptive capacity is not considered to be such a limiting factor, the schedule of the rate of return on capital is presumed to decline rather slowly.

FIGURE 1

Absorptive Capacity

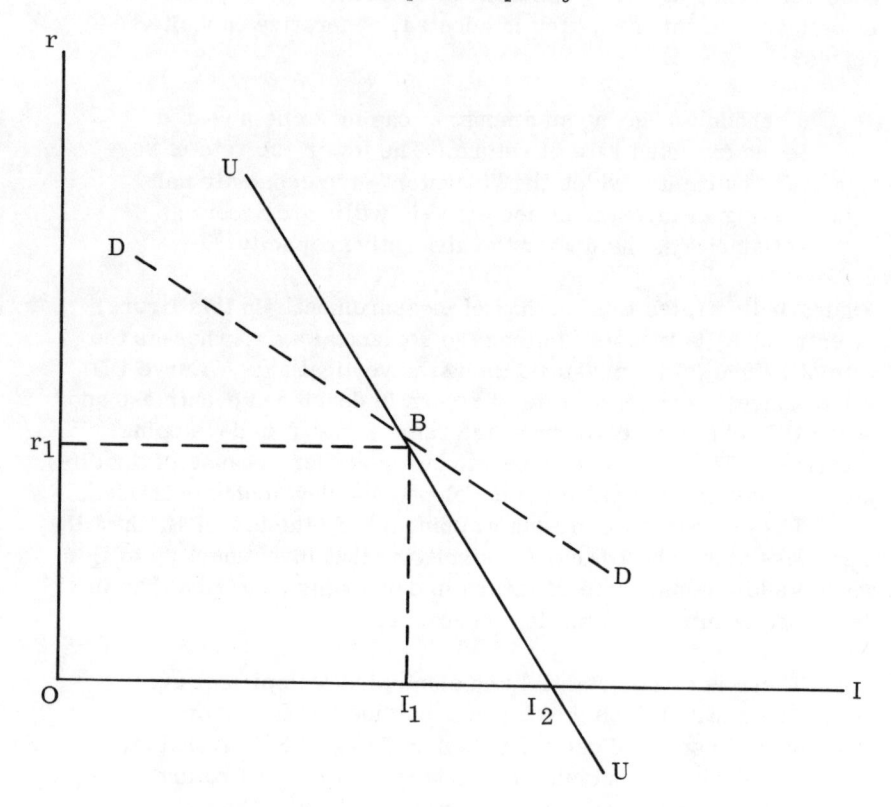

The preceding analysis relates to an earlier observation—namely, the absorptive capacity of the economy is that amount of invested capital that sets the rate of return exactly at the rate available outside the country. In Figure 1, the absorptive capacity is equal to 1_1. Any domestic investment beyond this point is inefficient.

It is, of course, entirely possible to ignore the rate of return (at least for a while) in developed countries. A country that possesses ample financial resources may deem it necessary, for social and political reasons, to expand domestic investment up to the point where the rate of return on the last unit of installed capital is equal to zero. In this case, the absorptive capacity of the economy becomes equal to I_2. Finally, one may conceive of the schedule relating investment to the rate of return in developing countries to possess a pronounced kink, that is, to assume the shape depicted by UBI_1. The kink could occur, for example, upon the exhaustion of a factor of production. If the country is unable to import this factor, further investment would be physically impossible. Furthermore, the cost of importing the factor in short supply could be so prohibitively high that the rate of return on investment would suddenly drop to zero or become negative. Under these circumstances the absorptive capacity of the economy would be equal to I_1.

Up to this point, the analysis has been based on the concept of a financial rate of return. The difficulties posed by the financial rate of return were cited earlier. To avoid them, it is possible to regard the vertical axis in Figure 1 as measuring the social rate of return. Assuming r_1 to be the "acceptable" social rate of return, then the absorptive capacity of the economy will again be equal to I_1. However, the concept of social rate of return is a subjective notion, a fact that demonstrates the difficulty of objectively measuring absorptive capacity.

In practice, absorptive capacity has been measured indirectly. For example, Chenery and MacEwan believe that "The most convenient measure of this absorptive capacity limit is the rate of increase in investment which a country can achieve on a sustained basis."[9] The most common rate of growth in investment observed by H. B. Chenery and A. MacEwan was between 15 and 20 percent per year.[10]

In addition to being comprehensible and unambiguous, this indirect method of measurement has the added advantage of focusing on the time dimension of absorptive capacity. It is useful to recognize that:

> The absorptive capacity of an economy depends on the time that is allowed for adjustments in the factors determining the limits. The more time is allowed to overcome the lack, or inadequate supply, of the co-operant factors, the greater the absorptive capacity becomes.[11]

It is likely that the passage of time will allow the process of development to increase the supply of cooperant factors in short supply and ameliorate intersectoral bottlenecks.[12]

This indirect method of measuring absorptive capacity hinges on the availability of investable funds. It is permissible to measure the absorptive capacity of the economy in this fashion if it can be proved that the economy is not handicapped by a savings-investment gap and an export-import gap. In the absence of this proof, it is not correct to measure the absorptive capacity of an economy on the basis of the observed rate of growth in investment. In recognition of this essential condition, Chenery and MacEwan observed that:

> The availability of external capital permits an economy to grow at the limit corresponding to its ability to increase its capital stock rather than at the lower rate implied by its ability to increase domestic savings.[13]

Limitations

In explaining the gradual nature of the development process, the World Bank listed a host of factors that limit the absorptive capacity of developing countries. These factors include the low level of education, frequent changes in government, and strong vested interests opposed to change.[14] It then concluded that "the principal limitation upon Bank financing in the development field has not been lack of money but lack of well-prepared and well-planned projects ready for immediate execution."[15]

It is not possible to have a complete catalog of limits that is applicable to each and every possible case. This is so because developing countries differ with respect to resource endowment. That is to say, limitations on absorptive capacity will necessarily vary from one country to another and, therefore, manifest themselves in many ways. As Adler points out, "The only way to come to grips with the practical limitations of absorptive capacity is to devise _specific_ measures to raise _specific_ limitations."[16]

Be that as it may, it is possible to categorize the various limits on absorptive capacity according to the broad classes of cooperant factors that could conceivably be in short supply.[17] Bearing this in mind, the following is a brief commentary on the most commonly observed limitations.

Limited Knowledge

A decisive factor that hampers the ability of developing countries to present well-conceived and ready-to-execute projects is their limited knowledge about the natural resources they possess. Adequate information about minerals, water resources, soils, and other natural endowments and climatic conditions is essential for the preparation of economically viable projects.

Scarcity of Entrepreneurs

Scarcity of entrepreneurial talents derives from Albert O. Hirschman's notion of "ability to invest."[18] Thus, according to one source:

> Such lack of "absorptive capacity," which essentially reflects an inability to invest in soundly conceived development programs and projects that can be carried out well and operated economically upon completion, is a common characteristic of less developed countries.[19]

Evidently, the availability of investment opportunities, such as the development of natural resources, is not by itself sufficient. There must exist a class of entrepreneurs ready and able to take advantage of these opportunities. According to the proponent of the unbalanced growth doctrine, developing countries normally suffer from the scarcity of indigenous entrepreneurs, a fact that constitutes a limit on their absorptive capacity.

Scarcity of Skills

Engineers, financial experts, executives, and skilled workers are all needed for the design, execution, operation, and maintenance of projects. Yet, underdevelopment implies, among other things, the scarcity of skills. This structural problem is recognized to be one of the most restrictive limitations. Drawing on the Nigerian experience, Wolfgang F. Stolper states:

> Absorptive capacity is therefore limited by the structure of the economy and the fact that investment decisions must be made over time. There are, of course, other limiting factors. Executive personnel are scarce and lose their effectiveness when overworked. Lack of executive capacity is a further limitation on absorptive capacity.[20]

Lack of skilled management is thought to be a most serious bottleneck for publicly owned enterprises. "The absence of efficient management is of particular importance in the case of state enterprises which for political reasons find it difficult to accept foreign management."[21]

Institutional Limitations

Institutional limitations are those that cannot be ameliorated at the project level. They are weaknesses in the institutional structure of society. Examples include inability to maintain order, uphold the laws, and protect property; excessive bureaucratic controls; and cumbersome procedures in the areas of export and import licensing, permits to establish enterprises, foreign exchange allocation, repatriation of profits, and so on. In general, institutional limitations of this type lower the rate of return on investment and thus restrict the absorptive capacity of the economy.

Cultural, Social, and Political Limitations

Developing societies do not generally abide by the same set of cultural and social values adhered to in developed countries. Lack of interest in material wealth, desire to cling to the simple life, rejection of the rigid discipline of factory life, and the reluctance to follow a specific schedule of work are examples of the differences in outlook and attitude. Furthermore, religious beliefs, social customs, and superstitions usually have a retarding effect on the developmental efforts of developing countries.

Most damaging, of course, are the existence of vested interests opposed to change and lack of political stability. As was seen in the case of Iraq, political instability had a definite retarding impact on the implementation of economic programs and plans. Finally, political instability is one of the factors responsible for the flight of indigenous capital and, most important, for the migration of professionals. The so-called "brain drain" cannot but constitute a limit on the absorptive capacity of most developing countries. (There are exceptional cases where the migration of highly trained persons could be attributable to lack of employment opportunities in the home country.)

These are, then, the limitations on absorptive capacity. They generally fall into two categories: those that are amenable to foreign technical assistance and those that are time-dependent. For example, it is possible with the help of foreign experts to survey the country's natural resources. Foreign engineers, skilled workers, and managers can design, construct, and manage factories. On the other hand, foreign assistance cannot be counted on to change customs and attitudes or bring about political stability. That is to say, in the final analysis, expansion of absorptive capacity is an evolutionary process. In this connection, the strategy of economic development can play a decisive role. Comprehensive planning that is aimed at educating the general public and providing training tailored to the specific needs of the economy and designed to solve acute bottlenecks should lead, in time, to an expansion of absorptive capacity. In contrast, haphazard planning or no planning at all will undoubtedly help to perpetuate the existing state of affairs.

THE CASE OF IRAQ

Gross Domestic Fixed Capital Formation

Tables 5.1 and 5.2 present gross domestic fixed capital formation (GDFC) in Iraq during the 1953-73 period at current and constant 1969 prices, respectively. As can be seen in Table 5.2, real domestic investment rose from ID 101.4 million in 1953 to ID 269.2 million in 1973, reflecting improvement in the country's ability to invest. In relative terms, however, real investment as a proportion of real gross domestic product declined from 24.9 percent in 1953 to 19.8 percent in 1973. As can be ascertained from Table 5.2, real gross domestic fixed capital formation amounted, on average, to 22.2 percent of real gross domestic product during the ten-year 1953-62 period; the ratio declined to 15.5 percent during the following ten-year period. This explains the high rates of economic growth during the 1950s and the low rates of growth during the 1960s.

The distribution of gross domestic fixed capital formation according to economic sector is presented in Tables D.4 and D.5; the former contains data measured in current prices and the latter contains the same data measured in constant 1969 prices. These data also provide an explanation of the uneven growth of the Iraqi economy that was commented upon in the preceding chapter. The following facts illustrate the point. The oil sector is the largest sector in the Iraqi economy; its contribution to GDP has been at least 30 percent throughout the 1953-73 period. The agricultural sector is the second largest commodity producing sector after oil. Manufacturing, the third largest commodity producing sector, was until recently a very poor third. Yet the average annual real gross domestic fixed capital formation in these sectors during the 1953-73 period were ID 9.6 million, ID 17.6 million, and ID 26.9 million, in that order. Now, surely these figures are behind the fast growth of manufacturing and the slow growth of the oil and agricultural sectors. Similarly, the bottlenecks in construction that surfaced in 1974 and beyond can be explained by the fact that the average annual investment in this sector was no more than ID 2.7 million during this period.

Sectoral distribution of investment notwithstanding, it must be recognized that the overall performance of the Iraqi economy has

TABLE 5.1

Iraq: Gross Domestic Fixed Capital Formation at Current Prices, 1953-73
(millions of Iraqi dinars)

Year	Public Sector	Private Sector	Total	GDFC as Percent of GDP*
1953	40.3	41.9	82.2	25.4
1954	42.6	44.4	87.0	23.2
1955	45.0	46.8	91.8	23.7
1956	47.3	49.3	96.6	22.5
1957	58.8	47.5	106.3	24.7
1958	55.2	42.6	97.8	20.2
1959	52.4	51.2	103.6	20.3
1960	51.9	68.4	120.3	21.3
1961	59.6	77.6	137.2	22.3
1962	58.1	61.1	119.2	18.1
1963	56.8	50.6	107.4	16.0
1964	67.1	55.0	122.1	16.0
1965	71.2	58.6	129.8	15.0
1966	76.2	73.4	149.6	15.9
1967	79.7	64.0	143.7	15.2
1968	75.8	67.2	143.0	13.4
1969	78.4	78.8	157.2	14.2
1970	101.1	84.0	185.1	15.4
1971	105.0	89.7	194.7	14.2
1972	114.6	102.4	217.0	15.6
1973	218.9	69.7	288.6	18.2

*Calculated from data in this table and Table D.2.

Sources: Jawad Hashim, Hussein Omar, and Ali al-Munoufy, Evaluation of Economic Growth in Iraq, 1950-1970, vol. 1, The Planning Experience (Baghdad, Iraq: Ministry of Planning, 1970); Jawad Hashim, Fixed Capital Formation in Iraq: 1957-1970 (Baghdad, Iraq: Ministry of Planning, 1972); Iraq, CSO, Annual Abstract of Statistics 1974 (1957-1970); and Iraq, CSO, Annual Abstract of Statistics 1976.

TABLE 5.2

Iraq: Gross Domestic Fixed Capital Formation
at Constant 1969 Prices, 1953-73
(millions of Iraqi dinars)

Year	Public Sector	Private Sector	Total	GDFC as Percent of GDP*
1953	49.7	51.7	101.4	24.9
1954	55.6	57.8	113.4	22.7
1955	53.3	55.5	108.8	23.2
1956	54.2	56.4	110.6	22.0
1957	70.1	57.8	127.9	25.7
1958	67.0	52.3	119.3	21.6
1959	59.0	57.7	116.7	20.7
1960	56.4	75.7	132.1	21.1
1961	66.0	86.7	152.7	21.7
1962	65.0	68.4	133.4	18.0
1963	63.9	56.8	120.7	16.6
1964	72.8	59.1	131.9	15.7
1965	77.6	63.5	141.1	14.8
1966	80.7	76.2	156.9	15.8
1967	83.6	64.3	147.9	16.1
1968	77.4	67.7	145.1	13.4
1969	78.4	78.8	157.2	14.2
1970	98.5	81.8	180.3	16.1
1971	98.3	86.2	184.5	15.5
1972	109.0	96.7	205.7	16.7
1973	204.2	65.0	269.2	19.8

*Calculated from data in this table and Table D.3.
Sources: Jawad Hashim, Hussein Omar, and Ali al-Munoufy, Evaluation of Economic Growth in Iraq, 1950-1970, vol. 1, The Planning Experience (Baghdad, Iraq: Ministry of Planning, 1970); Jawad Hashim, Fixed Capital Formation in Iraq: 1957-1970 (Baghdad, Iraq: Ministry of Planning, 1972); Iraq, CSO, Annual Abstract of Statistics 1974 (1957-1970); and Iraq, CSO, Annual Abstract of Statistics 1976.

been reasonably good. Whether measured in current or constant prices, domestic capital formation was, on average, about 20 percent of GDP during the 1953-73 period. It also must be recognized that in the absence of absorptive capacity constraints the ratio of GDFC to GDP would have been higher. This is so because, as will be explained shortly, Iraq had a surplus of investable funds at its disposal throughout this period.

Definition of Absorptive Capacity

For the purpose of this study, Iraq's absorptive capacity in a given period (year) is defined as that level of gross domestic investment that the country achieved during the sample 1953-73 period. Specifically, it is postulated that Iraq's observed levels of gross domestic investment were the highest the country could undertake, given its absorptive capacity. The reasons underlying this assumption were touched upon in the preceding chapters and will be dealt with further in this section.

If this is the case, then it is possible to measure Iraq's absorptive capacity by ascertaining the rate of increase in gross domestic investment that the country achieved on a sustained basis. This method of measurement implies that the maximum level of gross domestic investment in any one period cannot exceed the level of gross domestic investment in the preceding period times an exogenously determined rate of increase. Symbolically, this relationship is expressed as follows:

$$I_t \leq I_{(t-1)} + BI_{(t-1)}$$

where

I_t = investment in period t
B = the exogenously determined rate of increase

The level of B is governed by the limits of absorptive capacity.

As indicated earlier, this method of measurement can only be valid if it can be proved that the economy was not hampered by a savings-investment and/or an export-import gap. Therefore, it is necessary to show that the Iraqi economy was not constrained by these two gaps. In this connection, our point of departure will be the following familiar identity:

$$I \equiv S \equiv S_d + (M-X)$$

where

    I   = gross investment
    S   = gross savings
    S_d = domestic savings
    (M-X) = foreign savings (the difference between imports and exports)

The above identity can be rearranged as follows:

$$I - S_d \equiv M - X$$

In most developing countries domestic savings fall short of investment, and the balance is provided by foreign sources and expressed as import surplus (M > X). That is to say, investment in most developing countries is constrained by the savings-investment and the export-import gaps.[22] The reverse case occurs when the import surplus changes into an import shortage, that is, where X > M. In other words, the usual savings-investment and export-import gaps are reversed.

In the familiar case of most developing countries I > S_d and M > X, that is, both sides of the identity are positive. In this situation it is not correct to measure the absorptive capacity of the economy on the basis of the observed rate of increase in gross domestic investment unless it can be shown that capital inflows did in fact exceed the magnitude of the gap on a consistent basis. On the other hand, it is perfectly legitimate to measure the absorptive capacity of the economy in this manner if it can be shown that I < S_d and M < X—that is, if it can be shown that both sides of the identity are negative. When domestic savings exceed domestic investment, the excess will manifest itself at the beginning as a current account surplus. That is to say, when S_d > I, then X > M. The country involved will undertake investment in foreign countries; it is a capital exporter.

Measurement of Absorptive Capacity

After this digression, it is time to examine the case of Iraq. The fact was brought out in Chapter 2 that in the span of 25 years Iraq experienced a current account deficit in only 6 years. These deficits were caused not by a surge of imports but rather by a shortfall in exports due to external factors such as wars in the Middle East or decisions by foreign companies operating in Iraq to reduce crude oil production and exports. Between 1950 and 1974 Iraq realized a cumulative current account surplus amounting to ID 1,346.4 million (Table D.1). The bulk of this surplus occurred in later years, but this only demonstrates that after more than two decades the Iraqi economy was still unable to surmount some of the limitations on its absorptive capacity.

The cumulative current account surplus is nothing but a savings-investment and export-import gap in reverse. It amounted to 77 percent of total public sector investment expenditures during the 1951-74 period and 40 percent of gross domestic fixed capital formation during the 1953-73 period. These are significant magnitudes. In the absence of absorptive capacity limitations Iraq would have been able to achieve a grand total of gross domestic fixed capital formation equivalent to ID 4,758.5 million instead of the ID 3,412.1 million it actually achieved during the 1953-74 period (in current prices).

It is possible to look at the limits of the absorptive capacity of the Iraqi economy from a different perspective. The following manifestations of absorptive capacity constraints were discussed in Chapters 3 and 4.

1. Planned public sector investment expenditures consistently exceeded actual expenditures. In fact, the overall rate of implementation between 1951 and 1974 was no more than 56 percent.

2. The shortfall in actual expenditures was not caused by lack of funds. Between 1951 and 1974, revenues accruing to the development budget exceeded actual expenditures by ID 366 million.[23] Put differently, the actual surplus amounted to 25 percent of actual expenditures.

3. The shortfall in actual expenditures was due to delays in project execution because of shortage in skills, cumbersome government routine, and so on.

4. The government neither mobilized domestic private savings for development purposes nor encouraged private investment in a meaningful manner.

5. The government did not marshall foreign savings nor encourage foreign private capital inflows.

6. No significant effort was exerted to obtain official foreign loans. Iraq's foreign debt amounted, in March 1976, to ID 128.5 million.[24] This small amount was not contracted because the government felt the need for funds; rather they were politically motivated. The Soviet and Czech loans (amounting together to ID 77.2

million) were contracted in the early 1960s in order to strengthen cooperation with the socialist countries.

7. The inability to invest was not caused by a dearth of ready projects. Some of the projects that were included in the general programs of the DB during the 1950s were not acted upon until the 1960s and 1970s.

Having established that the Iraqi economy has not been constrained by lack of savings or foreign exchange, it is now possible to measure its absorptive capacity according to the proposed method, that is, by estimating the rate of increase in investment. Data on Iraq's real gross domestic fixed capital formation during the 1953-73 period are presented in Table 5.2 and Table D.5. In Table 5.2 the data are broken down according to the private and public sectors; in Table D.5 they are broken down according to the various economic sectors.

These data were used to measure the absorptive capacity of the Iraqi economy. Data in Table 5.2 were employed to measure the general absorptive capacity—that is, the rate of increase in capital formation in the economy as a whole and the rates of increase in both the public and private sectors. Data in Table D.5 were similarly used to estimate the absorptive capacities of the agricultural and industrial (manufacturing, construction, and electricity and water) sectors. The calculated rates of growth are presented in Table 5.3.

Two aspects required careful consideration before any calculation was actually made. These are the method of estimation and the time span. As for method, it is possible to calculate a rate of growth in several ways. An annual average rate can be calculated by comparing the gross domestic fixed capital formation in 1973 with the gross domestic fixed capital formation in 1953. Alternatively, it would be possible to calculate the rate of growth in every year and take the average (mean) as the representative rate of growth during the period in question. There are other ways. For example, a growth rate can be calculated by comparing the average of gross domestic fixed capital formation during the 1971-73 period with the average of gross domestic fixed capital formation during the 1953-55 period.

However, all of the above methods cannot be regarded as satisfactory because they involve a comparison of two points in time and ignore developments during intervening years. It is entirely possible that Iraq had several recessions and/or booms during the period under investigation, but the above methods would have completely missed the impact of such possible developments. Another source of bias is the state of the economy during the base

TABLE 5.3

Aggregate and Sectoral Growth Rates of Gross Domestic Fixed Capital Formation in Iraq
(percent)

Sector/Period	1954-73	1964-73	1954-58	1959-63*	1964-68	1969-73
Total GDFC	3.47	6.50	3.46	n.s.	4.00	11.85
GDFC in the private sector	2.08	3.62	n.s.	n.s.	3.82	n.s.
GDFC in the public sector	4.46	8.00	6.50	n.s.	4.12	18.15
GDFC in agriculture	3.33	12.82	10.23	-9.54	13.38	12.64
GDFC in industry	6.78	7.20	9.41	n.s.	8.37	9.38

n.s.: not significant.
*Period of political instability.
Source: Table D.7.

year and the end year. If the economy were booming during the base year and depressed during the end year, then the rate of growth would be underestimated, and vice versa.

These considerations dictated the choice of the "ordinary least squares" methodology (see Appendix B). This method of calculation is assumed to show a trend over time rather than the growth rate between two points in time. Put differently, this method is assumed to choose the growth rate over time that "fits" best with all of the varying annual growth rates. Since the objective is to ascertain that rate of increase in investment that Iraq "can achieve on a sustained basis," the least squares estimation is undoubtedly the most appropriate procedure.

The desirable properties of the least squares technique are well known and need not be repeated here. Suffice it to say that the validity of the results obtained with this technique is established by such tests as confidence intervals and correlation. Indeed, these checks were relied upon to single out (with an asterisk in Table D.7) those calculated growth rates that were statistically not significant.

The other aspect that required careful consideration was the time span over which the rate of growth was to be ascertained. Recalling earlier discussion about the time dimension of absorptive capacity, it is obvious that a 20-year span could underestimate the sustainable rate of growth of investment that Iraq could achieve. On the other hand, it can be argued that a 5-year period is too short, and, therefore, a growth rate so calculated cannot be regarded as a trend unless it is clearly a part of a pattern that is sustainable across sectors.

To highlight these conceptual difficulties growth rates were calculated over a 20-year period, a 10-year period, and a 5-year period. The results confirmed our intuition. As can be ascertained from Table 5.3, rates of growth over the 1954-73 period were lower than the rates of growth over the 1964-73 period. Moreover, due to the improvement in absorptive capacity of the economy over time, the rates of growth over the 1954-58 period were lower than the rates of growth over the 1969-73 period.

The results also confirm the role of political instability as a limiting factor. The 1959-63 period was very unstable politically in the modern history of Iraq. Hence, the rates of growth over this period were generally not statistically significant. The only exception was in agriculture where the rate of growth was negative (-9.54). This is a reflection of the fact that during the 1959-64 period, the authorities were preoccupied with industrialization to the neglect of agriculture (see Chapter 2).

The results also confirm the observation stated in the preceding chapter concerning the neglect of the private sector. Note that the rates of growth of GDFC in this sector were not statistically insignificant during the 1954-58, 1959-63, and 1969-73 periods. Note also that the rate of growth was a little over 2 percent during 1954-73 as compared with 3.5 percent for the economy as a whole and 4.5 percent for the public sector. Of special interest is the fact that the rate of growth of GDFC in the private sector during 1964-73 was 3.6 percent. This means that the rate of growth in this sector during 1954-63 was much less than 2 percent. This is paradoxical. During the 1950s, Iraq was supposedly operating under the free enterprise system. Logically, private investment should have been more depressed during the 1960s owing to political instability and the nationalization measures of 1964 and as a natural consequence of the transformation into a socialistic economic system. The fact that the reverse was true is further proof that the charge that the Development Board neglected the private sector was richly deserved.

It is not always possible to draw a positive relationship between the rates of growth of investment and GDP. This is at least true in the case of Iraq. Examination of Tables 4.1 and 5.3 illustrates the point. Total GDFC grew at the rate of 3.5 percent per annum during 1954-73, whereas the rate of growth of GDP was 6 percent. This is somewhat surprising but not unexpected. As indicated earlier, GDFC as a proportion of GDP has declined over time (Table 5.2). More puzzling is the fact that during 1964-73 total GDFC grew at an annual rate of 6.5 percent, while the rate of growth of GDP declined to 4.7 percent. It appears that GDFC was proceeding at a higher rate of growth precisely during the period when its share in GDP was declining and that the growth of GDP was proceeding at a lower rate precisely during the period when GDFC was growing at a fast rate.

These seemingly paradoxical movements can readily be explained. In regard to investment, it must be noted that during the mid-1960s investment was very depressed. The rate of growth in GDFC during 1964-68 was no more than 4 percent. Substantially higher rates were achieved in later years (11.85 percent during 1969-73). It is during these latter years that the share of GDFC in GDP began to rise. The higher rates in 1969-73 caused the overall rate for 1964-73 to be significantly higher than the one observed over the 20-year 1954-73 period.

In regard to GDP, the explanation is a bit more complex. Growth of GDP in Iraq has been led by the distribution and services sectors. Over the 1954-73 period the annual rate of growth was 6.9 percent in the distribution sectors and 9 percent in the services sectors. In contrast, the rate of growth was 4.5 percent in the agricultural sector, 4.7 percent in the mining and quarrying (oil) sector, and 6.7 percent in the industrial sector. Though the rate of growth of industrial output was reasonably high, its impact on GDP was limited, owing to the small size of this sector.

During the 1964-73 period, the growth rates of the distribution and services sectors dropped significantly to 4 percent and 6.5 percent, respectively. This fact accounts for the drop in the overall rate of growth to 4.7 percent, as against the 6 percent achieved over the 20-year period of 1954-73. An equally important factor behind this reversal has been the direction of the investment effort. Despite the fact that investment during the 1960s proceeded at a much higher rate than in the 1950s, its impact on GDP was limited. This is so because investment was concentrated in physical and human infrastructural projects—projects with long gestation periods, whose immediate contribution to output could only be very limited. This is clearly evident in the case of agriculture. Note that investment in the agricultural sector grew at an annual rate of 10.2 percent during 1964-73, but the growth of agricultural output was only 2.8 percent over the same period. On the other hand, the high rate of investment in industry has resulted in a high rate of growth in industrial output. But, once again, this sector is too small to make a noticeable difference in the overall rate of growth.

Given its social priorities and the modest state of its economic base, Iraq may have had little control over the course of economic events in the past. Following the profit-sharing agreement of 1952, oil revenues increased substantially and continued to do so at an accelerating rate. In such circumstances, it is expected that the government would expand its social programs, especially in the areas of health, education, and housing. This explains the high rates of growth in the services and distribution sectors.

With respect to the direction of development expenditures, the country had little choice, at least during the early years, but to concentrate on building the necessary infrastructure before turning its attention to productive projects. Industrial investment, by its nature, has been the most notable exception where from the very beginning the focus has been on import substitution. This accounts for the vigorous growth of industrial output.

The pattern of development expenditures in the past should have a favorable impact in the future. In the 1980s and beyond, the country should reap the benefits of its earlier investment in roads, bridges, water conservation, communications, and, most important, manpower resources. The coming years should witness less emphasis on infrastructure and more emphasis on production projects. And, hopefully, this will yield higher rates of economic growth.

NOTES

    1. Sayre P. Schatz, "The Capital Shortage Illusion: Government Lending in Nigeria," Oxford Economic Papers 17 (1965): 309.
    2. It was stated in the Fourth Annual Report of the IBRD that, "Perhaps the most striking single lesson which the Bank has learned in the course of its operations is how limited is the capacity of the underdeveloped countries to absorb capital quickly for really productive purposes." See IBRD, Fourth Annual Report to the Board of Governors: 1948-1949, p. 8.
    3. John H. Adler, Absorptive Capacity, the Concept and Its Determinants (Washington, D.C.: Brookings Institution, 1965), p. 1.
    4. Branko Horvat, "The Optimum Rate of Investment," The Economic Journal 68 (1958): 748.
    5. Ibid., p. 756.
    6. Adler, Absorptive Capacity, p. 5.
    7. Ibid., p. 2.
    8. Ibid., pp. 2-4.
    9. H. B. Chenery and A. MacEwan, "Optimal Patterns of Growth and Aid: The Case of Pakistan," in The Theory and Design of Economic Development, ed. Irma Adelman and Erik Thorbecke (Baltimore: Johns Hopkins University Press, 1966), p. 151.
    10. Ibid.
    11. Adler, Absorptive Capacity, p. 28.
    12. Hollis B. Chenery and Alan M. Strout, "Foreign Assistance and Economic Development," The American Economic Review 56 (1966): 680. There is no guarantee that time alone will lead necessarily to an expansion of the absorptive capacity of the economy. It all depends on the strategy of economic development. A strategy aimed at the development of human resources and easement of sectoral bottlenecks will certainly contribute to the expansion of absorptive capacity.
    13. Chenery and MacEwan, "Optimal Patterns of Growth and Aid," p. 152.
    14. IBRD, Fourth Annual Report, pp. 8-9.
    15. Ibid., p. 9.
    16. Adler, Absorptive Capacity, p. 31.
    17. Ibid., pp. 31-34. The remaining discussion in this section is mainly based on Adler's list of limitations.
    18. Albert O. Hirschman, The Strategy of Economic Development (New Haven, Conn.: Yale University Press, 1958).
    19. Albert Waterston, Development Planning: Lessons of Experience (Baltimore: Johns Hopkins University Press, 1965), p. 300.

20. Wolfgang F. Stolper, Planning without Facts, Lessons in Resource Allocation from Nigeria's Development (Cambridge, Mass.: Harvard University Press, 1966), p. 58.

21. Adler, Absorptive Capacity, p. 23.

22. For a concise statement of the theory of the two gaps see Jaroslav Vanek, Estimating Foreign Resource Needs for Economic Development (New York: McGraw-Hill, 1967), especially chap. 6.

23. Iraq, CBI, Bulletin, no. 1 (January-March 1976); and Iraq, CSO, Annual Abstract of Statistics 1976. Actual surplus is much higher than the figure indicated above. This is so because loans and grants to the current budget and public and semipublic institutions were regarded as development expenditures. For example, in 1959 loans amounting to ID 59.4 million were deemed as development expenditures.

24. Ibid.

6

THE IMPACT OF THE RISE IN OIL PRICES ON THE ABSORPTIVE CAPACITY OF THE IRAQI ECONOMY

The absorptive capacity of the Iraqi economy during the period before the rise in oil prices was defined and measured in the preceding chapter. The impact of the rise in oil prices on this absorptive capacity and the causes of this impact will now be examined.

ACTUAL VERSUS EXTRAPOLATED DOMESTIC INVESTMENT

It was indicated earlier that the most straightforward method of gauging the impact of the sharp increase in oil revenues on Iraq's absorptive capacity would be to compare extrapolated values of gross domestic capital formation with actual values during the years following the rise in oil prices.

Actual total gross domestic fixed capital formation and actual gross domestic fixed capital formation in both the private and public sectors for the 1974-76 period are presented in Table 6.1.[1] Table D.5 contains the same information according to economic sector for the same period.*

There are two observations to be made regarding the data in Table 6.1. First, the absolute level of GDFC more than tripled during the 1973-76 period. Second, the ratio of GDFC in 1969 prices to real GDP in prices of that year rose from 20 percent in

---

*It was not possible to obtain a breakdown of GDFC during 1976 according to economic sector.

1973 to 44 percent in 1976. The sharp increase is mainly attributable to the terms of trade income effect, an aspect that will be fully explored in the following section. Suffice it to say here that in the absence of this impact the ratio of GDFC to GDP during the 1974-76 period would have been very close to the level observed over the 1953-73 period.

TABLE 6.1

Iraq: Gross Domestic Fixed Capital Formation, 1974-76
(millions of Iraqi dinars)

	Current Prices			Constant 1969 Prices		
	1974[a]	1975[b]	1976[b]	1974[a]	1975[b]	1976[b]
Public sector	446.0	632.1	809.7	410.1	572.6	676.6
Private sector	85.9	129.1	239.0	79.1	116.9	196.3
Total	531.9	761.2	1,048.7	489.2	689.5	872.9
GDFC as percent of GDP[c]	15.9	19.2	22.9	32.6	38.8	43.8

[a] Revised.
[b] Provisional.
[c] Calculated from data in this table and Table D.3.

In extrapolating gross domestic fixed capital formation, it was necessary to choose the appropriate time span. As pointed out in Chapter 5, growth rates over the 1954-73 period were generally low, whereas growth rates over the 1969-73 period were generally very high. It is not difficult to argue in favor of rejecting both sets and accept the growth rates over the 1964-73 period as representative of secular trends in Iraq.

Be that as it may, two sets of extrapolations were calculated. The first set (Case I) is based on the growth rates over the 1964-73 period; and the second set (Case II) is based on the growth rates over the 1969-73 period. The results together with actual values are presented in Table 6.2. As can be seen in this table, actual real values exceed extrapolated values by substantial margins. In Case I, it is seen that in 1974 actual total GDFC exceeded extrapolated GDFC by more than 70 percent. The increase was even more pronounced in the public sector. Examination of GDFC in the various economic sectors reveals that the largest increase was in industry

TABLE 6.2

Iraq: Actual versus Extrapolated Gross Domestic Fixed Capital Formation at Constant 1969 Prices (millions of Iraqi dinars)

	1974			1975			1976		
	1	2	3	1	2	3	1	2	3
	Extrapolated	Actual	2 to 1 (percent)	Extrapolated	Actual	2 to 1 (percent)	Extrapolated	Actual	2 to 1 (percent)
Total GDFC									
Case I[a]	286.7[b]	489.2	170.6	305.3[b]	689.5	225.8	325.2[b]	872.9	268.4
Case II[c]	301.1	489.2	162.5	336.8	689.5	204.7	376.7	872.9	231.7
GDFC in the private sector									
Case I	67.4[b]	79.1	117.4	69.8[b]	116.9	167.5	72.4[b]	196.3	271.1
Case II	n.a.	79.1	n.a.	n.a.	116.9	n.a.	n.a.	196.3	n.a.
GDFC in the public sector									
Case I	220.5[b]	410.1	186.0	238.1[b]	572.6	240.5	257.2[b]	676.6	263.1
Case II	241.3	410.1	170.0	285.1	572.6	200.8	336.8	676.6	200.9
GDFC in agriculture									
Case I	35.6	43.8	123.0	40.2	45.5	113.2	45.3	n.a.	n.a.
Case II	35.6	43.8	123.0	40.1	45.5	113.5	45.2	n.a.	n.a.
GDFC in industry									
Case I	86.8	140.1	161.4	93.0	236.6	254.4	99.7	n.a.	n.a.
Case II	88.6	140.1	158.1	96.9	236.6	244.2	106.0	n.a.	n.a.

n.a.: Data not available.

[a] Base year is 1973; trend is over 1964-73 (col. 2 in Table 5.3).

[b] The sum of extrapolated GDFC in both the private and public sectors does not add up to total extrapolated GDFC because independent growth rates are used for these extrapolations.

[c] Base year is 1973; trend is over 1969-73 (col. 6 in Table 5.3).

Sources: Extrapolated values are calculated from data in Table 5.2, Table D.5, and Table 5.3. Actual values are from Table 6.1 and Table D.5.

and the smallest was in agriculture. Two factors account for this result. First, the rate of growth of investment in agriculture during 1964-73 was higher than the rate of growth of industrial investment, and thus it was difficult to achieve a substantial increase over such a high rate. Second, it is much easier to speed up the implementation of projects in industry than in agriculture. Under Case II, the margins were slightly lower; for example, total actual GDFC exceeded extrapolated GDFC by 62.5 percent. This is to be expected in view of the fact that extrapolations in Case II were based on growth rates over the 1969-73 period, which were higher than the growth rates observed over the 1964-73 period.

In 1975 the excess of actual over extrapolated values was much higher. Under Case I, actual total GDFC exceeded extrapolated GDFC by about 126 percent; under Case II, the excess was about 105 percent. With the exception of agriculture, the margins in all other sectors were of the same order of magnitude. The margin became even wider in 1976; for example, actual total GDFC exceeded extrapolated GDFC by 168.4 percent under Case I and by 131.7 percent under Case II. The most encouraging development in 1976 was the continuation of the rising trend in private sector investment. As shown in Table 6.1, actual private sector investment rose from ID 79.1 million in 1974 to ID 116.9 million in 1975 and to ID 196.3 million in 1976. This indicates that additional investment is profitable.

These results demonstrate that the rise in oil prices had a positive impact on the absorptive capacity of the Iraqi economy. The conclusion is tentative in nature because it rests on three sets of observations. Developments during 1974-76 do not constitute a trend, and, therefore, the absolute level of GDFC could conceivably revert to historical standards. It must be pointed out, however, that the possibility that investment expenditures would drop below the high levels of recent years is remote. This is so because the tests employed here are heavily biased against the hypothesis. The first source of bias stems from the fact that extrapolated values were based on actual domestic investment expenditures in 1973, which were extremely high in comparison with secular trends. For example, total GDFC grew in 1973 by about 31 percent as opposed to 6.5 percent over the 1964-73 period. The second source of bias stems from the fact that the highest observed growth rates, specifically those pertaining to the 1964-73 and 1969-73 periods were also used to extrapolate domestic investment in Iraq over the 1974-76 period.

Even with these biases actual levels of GDFC exceeded extrapolated values by substantial margins. Therefore, it is not unreasonable to conclude that, based on these results, the rise in oil prices had and will continue to have a positive impact on the absorptive capacity of the Iraqi economy. Of course, it is entirely possible

106 / OIL REVENUES & ACCELERATED GROWTH

that rates of growth in the future could be lower than those observed during 1974-76, but such developments would not diminish the significance of the conclusion. What matters here is that the investment function, and not its slope, had experienced a permanent upward shift. Put differently, the rise in oil prices had caused a discontinuity in this function. The next task is to explain the factors responsible for this expansion in absorptive capacity.

THE TERMS OF TRADE EFFECT

A country's terms of trade is defined as P_x/P_m, where P_x stands for the price of the country's exports and P_m stands for the price of its imports.[2] Depending on the movements of these two price indexes, a country may gain or lose on account of its trade. To illustrate the concept, consider a country that exports good X at price P_x and imports good Y at price P_m. Net receipts in year 1 can be calculated as follows:

$$R_1 = P_{x1}X_1 - P_{m1}Y_1 \tag{6.1}$$

Similarly, net receipts in year 2 can be calculated as follows:

$$R_2 = P_{x2}X_2 - P_{m2}Y_2 \tag{6.2}$$

The subscripts in equation (6.1) and equation (6.2) refer to years 1 and 2, respectively.

R_2 will be equal to R_1 when everything remains the same or when changes offset each other. Normally R_2 does not equal R_1. The difference between them is usually caused by a change in prices, a change in quantities, or both. That component of the difference that is attributable to changes in prices is defined as a gain (loss) from changes in terms of trade.[3] It is measured as follows:

$$G = R_2 - R'_2 \tag{6.3}$$

where

G = gains from changes in the terms of trade, and

$$R'_2 = P_{x1}X_2 - P_{m1}Y_2$$

Equation (6.3) states that the gain from a change in the terms of trade is equal to net receipts actually realized minus net receipts that would have been realized had the quantities of year 2 been traded at the prices of year 1.[4]

At this stage a question arises as to the relevance of the gain from a change in the terms of trade to the concept of absorptive capacity. The relevance will become clear, however, upon an examination of the impact of changes in the terms of trade on "real income." Real income "may be described as a measure of how well off the country is in the current year, taking into account the effect of changes in the terms of trade between the base year and the current year."[5] In a closed economy, real product (GDP) is equal to real income (GDY). This is not necessarily true in the case of an open economy.

In this connection, it is helpful to view exports as an indirect method of "producing" imports. Viewing exports as a way of producing imports makes it possible to distinguish between real product (real GDP) and real income (real GDY). The World Bank developed a procedure that can be used to calculate real GDY.[6] The procedure is based on treating exports as a component of real GDP and real GDY. As a component of real GDP, real exports are measured by deflating the value of exports in current prices by the export price index.

> In measuring real GDY, however, "real exports" are looked upon as equivalent to the imports (in base period prices) which can be purchased with a country's export earnings. The income effect from changes in the terms of trade thus is the difference between GDY and GDP, i.e., $(E/P_m - E/P_e)$.[7]

Symbolically, real income can be expressed as follows:

$$GDY_k = GDP_k + (E/P_m - E/P_e) \qquad (6.4)$$

$$GDY_k = GDP_k + TTE*$$

where

k = subscript indicating base period prices

E = value of exports measured in current prices

P_m = price index for imports

P_e = price index for exports

TTE = terms of trade effect in base period prices

---

*The formal derivation of this equation is presented in Appendix C.

Equation (6.4) states that a country's real income (GDY_k) is equal to its real product (GDP_k) plus the income effect of a change in its terms of trade. If the change in terms of trade is in favor of the country in question, then its real income will be greater than its real product, and vice versa.

After this lengthy but necessary digression, the impact of the change in Iraq's terms of trade on its real income following the increase in oil prices can be examined. Before doing so, however, it is necessary to clear up certain aspects related to this exercise.

1. Since oil constitutes about 98 percent of Iraq's exports, it is not necessary to include noncommodity flows. Therefore, only merchandise exports are considered for the purpose of adjusting Iraq's real product to reflect the income effect of the change in its terms of trade.[8]

2. Two price indexes are needed for the performance of this exercise: a price index for Iraq's exports and another one for its imports. There exists a price index for Iraq's exports that is highly reliable, owing to the full documentation of oil prices. However, it has not been possible to obtain or construct a price index for Iraq's imports. The price index for the exports of the industrial countries has been chosen as a proxy for Iraq's imports price index. This is justifiable in view of the fact that a very high percentage of Iraq's imports originates in industrial countries. The two price indexes are presented in Table 6.3. They are the same as those to be found in the IMF, *International Financial Statistics*. The only difference is that they are expressed in Iraqi dinars.*

3. The year 1969 was chosen as a base year. This is legitimate in view of the fact that all of the quantitative analyses in this study were based on data expressed in 1969 prices. Furthermore, 1969 appears to be a representative year of Iraq's terms of trade. This can be ascertained from Table 6.3.

Iraq's exports expressed in current prices were deflated by the price index of the industrial countries' exports in order to calculate Iraq's "capacity to import." These exports also were deflated by the country's exports price index so as to express them in

---

*Price indexes in the IMF, *International Financial Statistics* are expressed in U.S. dollars. In order to express these indexes in Iraqi dinars, it is necessary to account for the movements of the Iraqi dinar against the U.S. dollar. This was done in connection with the construction of Table 6.3.

TABLE 6.3

Export Price Indexes in Iraqi Dinars
(1969 = 100)

Year	Iraq	Industrial Countries
1969	100.0	100.0
1970	100.0	106.0
1971	130.3	111.0
1972	120.7	112.0
1973	129.1	122.0
1974	433.6	150.0
1975	463.3	168.0
1976	495.4	170.0

Source: IMF, International Financial Statistics, Data Fund.

TABLE 6.4

Iraq: Terms of Trade Income Effect
(millions of Iraqi dinars)

Year	Exports (current prices)	Exports as Capacity to Import (1969 prices)	Exports (1969 prices)	Income Effect
1969	379.4	379.4	379.4	0.0
1970	390.6	368.5	390.6	-22.1
1971	544.9	490.9	418.2	72.7
1972	448.3	400.3	371.4	28.9
1973	652.1	534.5	505.1	29.4
1974	2,054.4	1,369.6	473.8	895.8
1975	2,450.2	1,458.4	528.8	929.6
1976	2,610.8	1,535.8	527.0	1,008.8

Sources: Exports in current prices are from IMF, International Financial Statistics, December 1976 and February 1978; other entries were derived from exports in current prices and price indexes in Table 6.3.

constant 1969 prices. The difference between these two magnitudes in each period is the income effect of the change in Iraq's terms of trade. The results are presented in Table 6.4. As can be seen in this table, the income effect rose from a negative ID 22.1 million in 1970 to a positive ID 895.8 million in 1974, the first year following the increase in oil prices, and to ID 1,008.8 million in 1976.

In accordance with equation (6.4), these annual income effects were incorporated into Iraq's national income accounts by simply adding them to the values of Iraq's real gross domestic product. The results are presented in Table 6.5 where it can be seen that Iraq's real income rose from 98.0 percent of real product in 1970 to 159.8 percent in 1974, 152.4 percent in 1975, and 150.6 percent in 1976. In absolute terms, Iraq's real income rose from ID 1,098.1 million in 1970 to ID 3,003.5 million in 1976. Real income in 1974 was 72 percent higher than real income in 1973; and real income in 1976 was 162 percent higher than real income in 1973. In other words, with the help of improvements in its terms of trade, Iraq's real income more than doubled in three years' time.

TABLE 6.5

Iraq: Real Product and Real Income at
Constant 1969 Prices, 1969-76
(millions of Iraqi dinars)

Year	1 Real Product (GDP)	2 Terms of Trade Effect	3 Real Income (GDY)	4 3 to 1 (percent)
1969	1,109.7	-0.0	1,109.7	100.0
1970	1,120.2	-22.1	1,098.1	98.0
1971	1,191.9	72.7	1,264.6	106.1
1972	1,230.5	28.9	1,259.4	102.3
1973	1,360.0	29.4	1,389.4	102.2
1974	1,499.2	895.8	2,395.0	159.8
1975	1,774.5	929.6	2,704.1	152.4
1976	1,994.7	1,008.8	3,003.5	150.6

Sources: Table D.3 and Table 6.4.

These results explain the positive impact of the rise in oil prices on the absorptive capacity of Iraq. If it is accepted that savings (investment) is a function of "real income," then the excess of actual over extrapolated GDFC is more or less fully explained, because GDFC as a percentage of GDY does not differ much from GDFC as a percentage of GDP. Furthermore, if it is accepted that demand for investable funds is a derived demand, then the expansion in Iraq's absorptive capacity is more or less fully explained.

More specifically, the explosion of investment in Iraq can be explained in terms of the acceleration principle that was "first stated by Sir Roy Harrod in 1936 and subsequently expanded by Professors Hicks and Samuelson in 1939."[9] The acceleration theory of investment can be stated as follows:

$$I_t = \beta \; Y_{(t-1)} - Y_{(t-2)} \qquad (6.5)$$

where

I_t = investment in period t

β = capital-output ratio

$Y_{(t-1)} - Y_{(t-2)}$ = rate of change in income in the previous period.[10]

Since β is technologically determined and is thought to be constant over the short run, it is clear from equation (6.5) that the level of investment is uniquely determined by the rate of change in income. If it is assumed here that Y in equation (6.5) refers to real income, then it follows that the rate of investment would depend on the rate of change in real income. Real income in Iraq rose by 160 percent during the three-year period that ended in 1976. According to the acceleration theory, investment should have increased in the manner that, in fact, did materialize.

Viewing investment as a function of real income provides, therefore, an explanation of the sharp rise in investment in Iraq expressed as a proportion of GDP. As a percentage of real GDP, real investment in Iraq rose from 20 percent in 1973 to 33 percent in 1974, 39 percent in 1975, and 44 percent in 1976. However, expressed as a percentage of real income, real investment amounted to 19 percent in 1973, 20 percent in 1974, 25 percent in 1975, and 29 percent in 1976. The latter set of ratios is in line with historical performance.*

---

*It can be ascertained from Table 5.2 that between 1953 and 1973, real investment in Iraq amounted, on average, to 19 percent

It is, of course, recognized that the acceleration principle is
". . . based upon a number of strong assumptions. . . . one of the
implicit assumptions is that the supply of capital is infinitely elastic
and consequently plays no part in the determination of the rate of investment. Another implicit assumption is that the accelerator is not
an economic relationship, but a technically given datum."[11] Although
the restrictiveness of these assumptions is acknowledged on theoretical grounds, it is likely that these assumptions strong as they are,
do hold in the case of Iraq. For all practical purposes the supply of
capital can be regarded as infinitely elastic, owing to the large current account surplus Iraq has been realizing and is expected to realize in the medium run. As regards the accelerator, it should be
pointed out that oil revenues accrue to the government. Therefore,
business savings, lags in dividend payments, and other such factors
that tend to diminish the impact of the change in real income[12] do not
come into play in this instance.

In trying to understand what actually happened in Iraq, it must
be kept in mind that demand for investment is a derived one. The
substantial increase in real income caused a substantial increase in
aggregate demand. This development is depicted in Figure 2 where
the aggregate demand curve DD shifted to the right and assumed the
position D'D'.

To satisfy the surge in aggregate demand, Iraq expanded its
imports. The value of these imports rose from ID 270.3 million in
1973 to ID 700.1 million in 1974 and ID 1,244.7 million in 1975.[13]
The country's desire to increase its imports was constrained by the

---

of real GDP. In comparison, the average ratio of real investment
to real income was 25 percent during the 1974-76 period. The difference can be explained as follows. First, between 1953 and 1973
Iraq consistently experienced a negative terms of trade income effect. Therefore, real investment as a proportion of real income was
certainly higher than 19 percent. Second, prior to the nationalization of oil in 1972, a sizable portion of GDP was used to repatriate
the profits of the foreign oil companies. The fact that oil revenues
now accrue in their entirety to the government may have caused the
ratio of GDFC during recent years to be somewhat higher than the
historical ratio.

Measuring the ratio of GDFC to GDP in current prices—that is,
ignoring the terms of trade income effect—corroborates the preceding reasoning. Over the 1953-73 period, nominal investment
amounted, on average, to 19 percent of nominal GDP. It averaged
19 percent during the three-year 1974-76 period.

inadequacy of port, transportation, and storage facilities. Consequently, the gap between aggregate demand and aggregate supply became wide enough to give rise to inflationary pressures.

FIGURE 2

Impact of TTE on Aggregate Demand

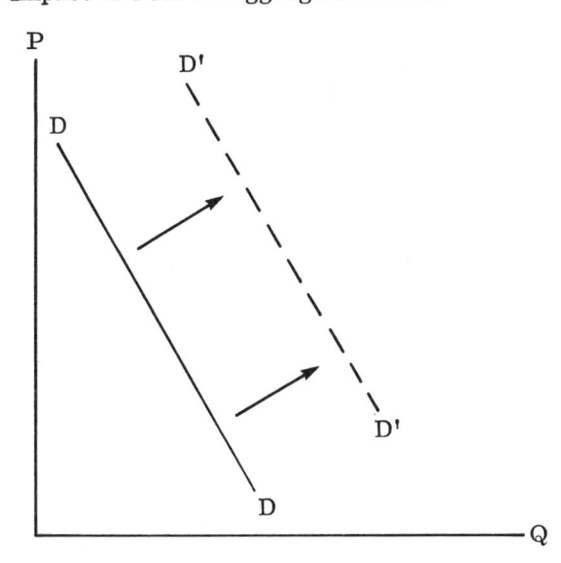

The implication of the rise in aggregate demand and prices is that it is profitable to increase investment at every level of acceptable rate of return. This fact in combination with the dedication of the authorities to the cause of development and their emphasis on import-substitution industrialization caused the schedule relating investment to the rate of return, curve UU in Figure 3, to shift to the right and assume the position indicated by U'U'. If it is assumed that the acceptable rate of return in Iraq is r_1, then its absorptive capacity has expanded from I_1 to I_2.

The dynamics of the rise in oil prices are such that a new interpretation of the concept of absorptive capacity is called for. Traditional theorizing about absorptive capacity is supply-oriented. The minimum acceptable rate of return is predicated on the idea that investable funds are in short supply. Not enough thought was given to the impact of a sufficient increase in aggregate demand on supply elasticities. Now, preliminary indications are that aggregate demand is a factor deserving of consideration in any analysis of absorptive capacity.

FIGURE 3

Impact of TTE on Absorptive Capacity

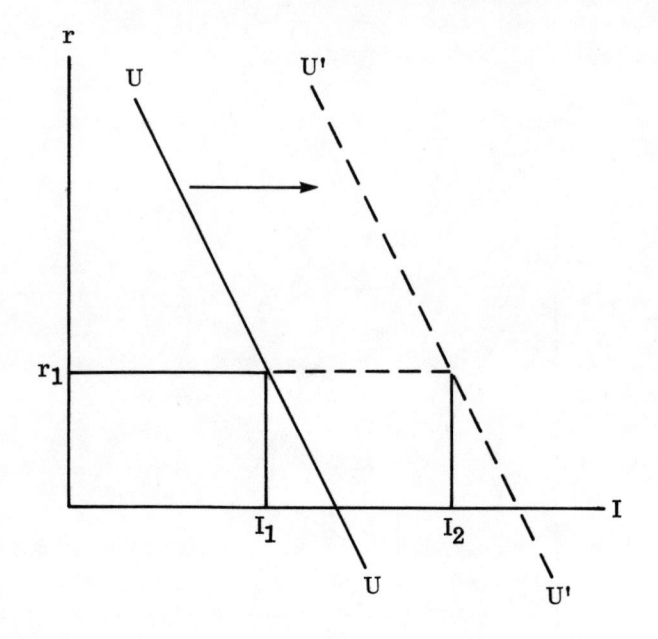

Before concluding this section it is useful to comment on a question that could be legitimately raised regarding the impact of the improvement in Iraq's terms of trade. Since it was shown in Chapter 5 that Iraq did not suffer from shortages in investable funds, the question, therefore, arises: Why should the availability of additional funds, owing to the terms of trade income effect, cause an expansion in the country's absorptive capacity? The answer can be summarized as follows:

1. It is a matter of degree. The financial resources that became available to the country after the rise in oil prices are extremely large in comparison with the financial resources that were at its disposal before the rise in oil prices. Thus, the impact of oil revenues on aggregate demand after the rise in oil prices is much greater than their impact before the rise in oil prices. Furthermore, the large sums of foreign exchange earnings in recent years has enabled the government to import most of the goods and services necessary for the implementation of development projects.

2. It is also a matter of confidence. Until 1973 the oil industry in Iraq was under the control of foreign oil companies. Considering the tense relationship that existed between the companies and

the Iraqi authorities, the government was more cautious in its spending policies. In contrast, the rise in oil prices came shortly after the nationalization of the oil industry. The combination of large revenues and knowledge that oil production and pricing policies would no longer be subjected to the wishes of the foreign oil companies gave momentum to the development effort.

OTHER PROBABLE CAUSES

Although it is concluded that the terms of trade income effect has been the prime mover behind the expansion of the absorptive capacity of the Iraqi economy, other factors relating to the significant rise in oil revenues also may have been important in this regard. It is not possible to quantify the impact of these factors. However, a fairly clear picture will emerge from the discussion that follows.

Foreign Labor

The acceleration in public and private expenditures caused a sudden and sharp increase in wages. The nominal wage of unskilled workers in Iraq rose more than threefold in 1974.* This is surely indicative of a shortage in labor supply. Aware of this situation, the government took measures to encourage Iraqi and Arab professionals and skilled workers in Europe and the United States to come to Iraq.[14] Foreign workers also were encouraged to work in the country.

Unfortunately, it has not been possible to gather data on Iraqi and Arab workers who returned to Iraq. However, their number is believed to be substantial. Inspection of the various issues of the Annual Abstract of Statistics published by the CSO reveals that the number of foreign workers in Iraq rose from 1,808 in 1973 to 2,523 in 1974.[15] It is not clear whether this number includes engineers and other professionals brought to Iraq by foreign contractors. Another difficulty stems from the fact that these figures do not include workers from Arab countries who are not regarded as foreign workers

---

*The Consumer Price Index in Iraq rose by 8.3 percent during 1974. Therefore, the rise in real wages was marginally less than the rise in nominal wages. See the following chapter on these points.

and who are believed to constitute the overwhelming majority of non-Iraqi workers in Iraq.

What holds true for foreign labor can be assumed to hold true for other cooperant factors. That is to say, the availability of oil revenues made it possible to augment the supply of cooperant factors through importation, a development that has been repeatedly stated as possible.

It was pointed out in Chapter 4 that shortages in skilled manpower have been Iraq's major bottleneck. Easing these shortages should, therefore, have a favorable impact on the country's absorptive capacity—that is, make additional investment over and above the level indicated by historical absorptive capacity profitable. And, indeed, further investment was undertaken.

Lower Rate of Return

It is not illogical to attribute the expansion in the absorptive capacity of the Iraqi economy to a decision by the authorities to accept a lower-than-normal financial rate of return. However, in the absence of clear-cut proof that the incremental capital-output ratio that prevailed during the 1974-76 period was higher than the incremental capital-output ratio that prevailed prior to the rise in oil prices, it would not be possible to confirm this hypothesis.

Indirect evidence suggests that this is not the case. Although it is reasonable to presume that public investment could be undertaken even though the financial rate of return is known to be lower than what is available outside the economy, it is not logical to so presume in the case of private investment. During the 1974-76 period, actual private investment exceeded extrapolated investment by substantial margins. As would be expected, the magnitude of the excess was much higher during 1975 and 1976 than in 1974. These results indirectly confirm that the financial rate of return on investment in Iraq has not declined. On the contrary, the increase in private investment expenditures indicates that the financial rate of return has been rising.

Be that as it may, it is entirely possible that the authorities decided to accept a lower financial rate of return. There are internal as well as external forces that are likely to encourage this course of action. The first factor to be considered is the opportunity cost of investable funds. Traditional theory of absorptive capacity is predicated on the assumption that investment funds in developing countries are in short supply. This is generally true except in the oil exporting countries. For this group of countries, the question of supply is no longer relevant. Their immediate problem is how to deal with the surplus of domestic savings.

To appreciate the problem of oil exporting countries, it is instructive to consider the problem of nonoil developing countries. In financing their domestic investment, nonoil developing countries first tap their domestic savings. When these savings are exhausted, these countries begin to tap foreign savings—that is, they begin to borrow. Borrowing implies that the financial rate of return on investment must be at least equal to the long-term rate of interest in the international capital markets. (The margin of financial intermediaries and other charges are ignored.)

The reverse is true in the case of the oil exporting countries. They can afford to invest in projects whose rate of return is equal to or less than the long-term rate of interest prevailing in the international capital markets. The reason that they can afford to do so is fairly obvious. Owing to the high rate of inflation in industrial countries and currency fluctuations, particularly the depreciation of the U.S. dollar, the oil exporting countries are finding it difficult to maintain the real value of their foreign holdings.

The next logical question is: Why should not these oil exporting countries place their surplus in direct investment where the rate of return is presumably higher? The question is legitimate. However, at this point, the second set of external factors manifests itself. In general, industrial countries find it difficult to allow the oil exporting countries to undertake direct investment in their countries. Direct investment in other developing countries is proceeding at a fast rate. However, it must be recognized that the oil exporting countries lack the necessary sophistication to undertake a great deal of direct investment. Furthermore, they face an uncertain future, particularly as regards nationalization, legislative impediments to profits repatriation, excessive taxation, and so forth.

Turning to internal factors, oil exporting countries must reconcile internal needs with the notion of efficient allocation of resources, viewed from a global point of view. The aspirations of the people cannot be ignored. The sudden and substantial increase in oil revenues has undoubtedly raised the expectations of the population. These expectations are reinforced by the dedication of the government to the cause of development. As far as Iraq is concerned, it is the official policy of the government to invest oil revenues domestically. The government's position was summed up by the director general of the Industrial Department in the Ministry of Planning, who was quoted as saying, "You will not see us buying stocks, putting money into banks, or buying islands off Carolina. . . . We are going to spend every penny right here."[16]

Another internal factor is shortage of labor. In these circumstances, the government may find it appropriate to substitute capital, which is in excess supply, for labor. That is to say, the government may find it appropriate to accept a high incremental capital-output ratio—that is, to accept a lower rate of return on invested capital.

Given their internal circumstances and their prospects abroad, it is not difficult to understand the decision of the authorities to lower their acceptable financial rate of return. Such a decision, however, does not necessarily imply a lower social rate of return. The social merits of additional investment could conceivably go far beyond the limited notion of profit.

Assume, for illustrative purposes, that the authorities decided to lower their rate of return from r_1 to r_2; then the absorptive capacity of the economy would have risen from I_1 to I_2 (Figure 4). At the margin, it is entirely possible that the government would accept a zero rate of return. Such a decision is justifiable on the grounds that the financial rate of return on foreign holdings could be negative. Under these circumstances, the absorptive capacity would be I_3.

FIGURE 4

Impact of a Lower Rate of Return on Absorptive Capacity

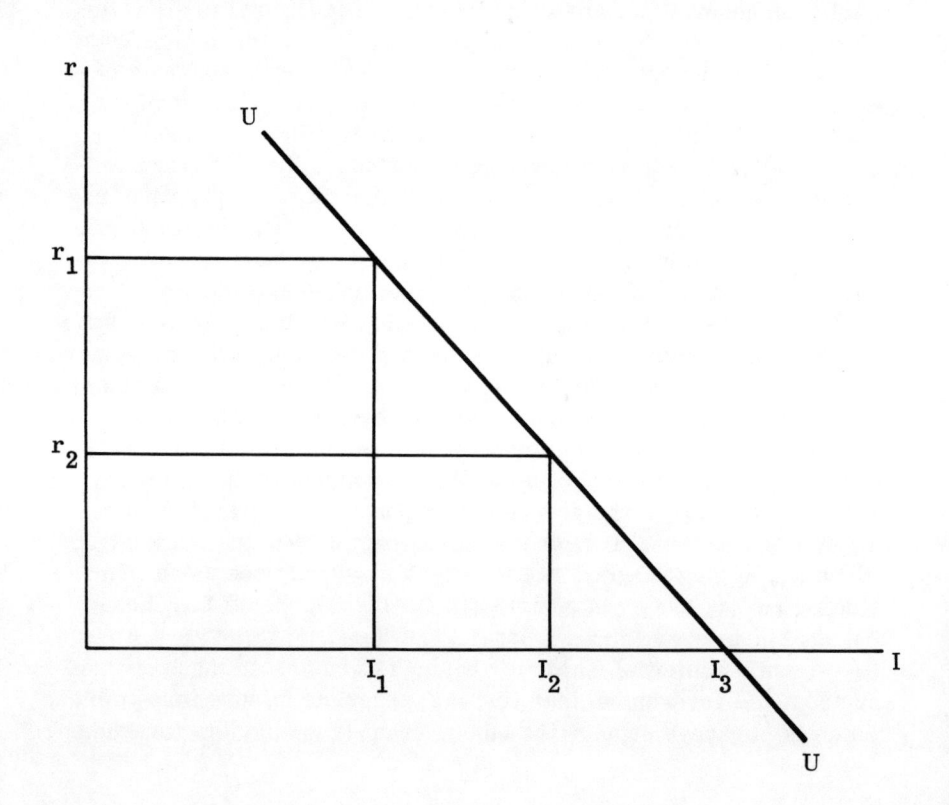

Intersectoral Impact

In Chapter 5, it was explained that bottlenecks in some sectors could be a cause of low absorptive capacity in other sectors. On the other hand, an increase in the demand of some sectors for the products or services of other sectors would raise the profits of the latter group and thus lead to an expansion in their absorptive capacities.

Although the expansion of the absorptive capacity of the commodities and services sectors is directly attributable to the substantial increase in real income, the expansion in the absorptive capacity of the distribution sectors is certainly attributable to the increase in the demand of the rest of the economy for their services. The nature of the impact is similar to the impact of the rise in real income on the overall absorptive capacity of the economy. In this respect, it is instructive to point out that investment in the transportation and communications sector in 1974 exceeded actual investment in 1973 by 142 percent, and the 1975 investment exceeded the 1973 level by 178 percent.

These results demonstrate the impact of the increase in the demand of other sectors for the services of the transport and communications sector. This fact was alluded to in Chapter 4 and the measures taken or planned were described briefly.

Sovereignty Over Natural Resources

Sovereignty over natural resources is a constraint that was not discussed in Chapter 5. Nevertheless, it is as real a constraint as the others, at least in the case of Iraq. Prior to nationalization, the Iraqi authorities were in no position to influence investment, production, and other policies connected with the oil sector.

The domination of this sector by the foreign oil companies was discussed in Chapter 2. Suffice it to say here that when the country was locked in a dispute with the foreign oil companies, investment in the oil sector suffered a great deal. Thus over the 1962-69 period, average annual GDFC in this sector was ID 1.9 million. Investment began to rise in 1970 and 1971, reflecting improved relations between the government and the foreign oil companies. It rose sharply following the nationalization of IPC on June 1, 1972. In 1973 GDFC in the mining and quarrying sector was ID 28.6 million, or 229 percent the level of GDFC in 1972 (Table D.5). Investment in this sector continued to increase, reaching ID 72.9 million in 1974 and ID 106.1 million in 1975.

National control over the oil sector helped the expansion in the absorptive capacity of the economy in yet another way. After nationalization the government became free to fulfill its stated policy of integrating the oil sector with the rest of the economy. Large-scale investment in refineries, petrochemicals, and other oil- and gas-based industries was undertaken, beginning in 1974. This explains the sharp increase in the value of GDFC in the industrial sector. Investment in the manufacturing subsector rose from ID 64.6 million in 1973 to ID 113.8 million in 1974 and to ID 195.8 million in 1975.

SUMMARY

This chapter began with a comparison between actual and extrapolated values of investment in Iraq. The fact that actual overall and sectoral investments substantially exceeded values indicated by the economy's absorptive capacity is strong evidence that the sharp rise in oil revenues had a positive impact on this absorptive capacity.

The decisive factor underlying this development is the large growth in Iraq's real income, which was a direct result of the improvement in the country's terms of trade. Other factors include the augmentation of the supply of cooperant factors through importation, the possibility that the Iraqi government decided to accept a lower financial rate of return, the intersectoral impact on the absorptive capacity of some sectors in the economy, and the decision of the government to nationalize the oil industry and integrate this industry with the rest of the economy.

Other decisive factors are the dedication of the government to the cause of development and the determination of the Iraqi people to develop their economy and achieve a better standard of living for themselves. The fact that the country has been enjoying political stability since 1968 has made this task so much easier. In the final analysis, political stability, strong leadership, dedication, and the passage of time will ease or completely eliminate the constraints that have limited the country's absorptive capacity in the past.

NOTES

1. Final figures of GDFC in Iraq provided by the Central Statistical Organization show that GDFC expressed in constant 1969 prices was ID 568.6 million during 1974 and ID 961.3 million during 1975. These figures are substantially higher than those presented in Table 6.1. Since it was not possible to obtain a breakdown of the new data according to economic sector, it was decided not to make use of these new figures; otherwise there would be a discrepancy

between figures presented in Table 6.1 and Table D.5. This decision is covered, however, by the notation in Table 6.1 that the figures presented therein are provisional. Even though the figures in Table 6.1 understate actual GDFC in Iraq during 1974 and 1975, they still demonstrate the fact that actual investment exceeded calculated investment by a wide margin. Finally, it should be noted that, according to the new data, GDFC during 1976 is lower than GDFC during 1975. This accords with the fact that the government decided, late in 1975, to slow down development and current expenditures in order to combat inflationary pressures and improve efficiency. This, however, was a temporary halt. The government resumed the acceleration of investment expenditures beginning in 1977. In this connection it should be pointed out that budgeted central government development expenditures amounted to ID 2,357 million in 1977, ID 2,800 million in 1978, and ID 3,283 million in 1979. See Middle East Economic Survey, March 28, 1977; and Middle East Economic Survey, April 16, 1979.

2. Bo Sodersten, International Economics (New York: Harper and Row, 1970), p. 26.

3. Anders Ølgaard, Growth, Productivity and Relative Prices (Amsterdam: North-Holland, 1966), p. 246. The gain from a change in the terms of trade may be defined in various ways, depending on assumptions made and price indexes used. Ølgaard established 12 different concepts of gain (see chap. 13 in this reference).

4. Ibid.

5. G. Stuvel, "Asset Revaluation and Terms-of-Trade Effects in the Framework of the National Accounts," The Economic Journal 64 (1959): 287.

6. IBRD, "The CPP System for Country Economic Projections," mimeographed (Washington, D.C.: World Bank, November 1973), chap. 6.

7. Ibid., p. 14.

8. It is not an easy task to deflate noncommodity flows because they do not have price indexes of their own. Normally, an acceptable proxy is used for this purpose. For some notable contributions in this field the reader is referred to Stuvel, "Asset Revaluation and Terms-of-Trade Effects"; J. B. Broderick, "National Accounts at Constant Prices," The Review of Income and Wealth 13 (1967): 247-58; Yoshimasa Kurabayashi, "The Impact of Changes in Terms of Trade on a System of National Accounts: An Attempted Synthesis," The Review of Income and Wealth 17 (1971): 285-97; and Raymond Courbis, "Comment on Y. Kurabayashi: The Impact of Changes in the Terms of Trade on a System of National Accounts," The Review of Income and Wealth 18 (1972): 247-50.

9. Joseph Aschheim and Ching-Yao Hsieh, *Macroeconomics: Income and Monetary Theory* (Columbus, Ohio: Charles E. Merrill, 1969), p. 116.

10. Ibid., p. 117, fn.

11. Aschheim and Hsieh, *Macroeconomics*, p. 118.

12. Ibid., p. 119.

13. IMF, *International Financial Statistics*, February 1978.

14. Republic of Iraq, *Public Law No. 154 of 1974*.

15. CSO, *Annual Abstract of Statistics 1973* and *Annual Abstract of Statistics 1974*.

16. *Business Week*, August 4, 1975, p. 34.

7

PROSPECTS FOR ACCELERATED ECONOMIC GROWTH IN IRAQ

This chapter consists of three parts. First, the prospects for accelerated economic growth in Iraq are assessed against the background of production and growth theories and the observed expansion in the country's absorptive capacity. The analysis is general; no attempt is made to specify the production function or the capital-output ratio in Iraq. However, the performance of the Iraqi economy during the 1974-76 period is compared with its performance during the sample 1953-73 period. Such a comparison should indicate whether the country has entered a period of accelerated growth.

Second, the conditions necessary to sustain accelerated growth are analyzed. The analysis includes policy measures, contemplated or already implemented, to ease the constraints that have limited the country's absorptive capacity in the past. The question of oil prices and the impact of future developments in this field on Iraq's terms of trade also are addressed.

Finally, plans to further develop the oil sector in Iraq and the implication of these plans for the rest of the economy are examined briefly.

ABSORPTIVE CAPACITY AND ACCELERATED GROWTH

Theoretical Discussion

The acceleration principle was cited in Chapter 6 to explain the expansion in the absorptive capacity of the Iraqi economy. In this chapter another principle, namely, the functional relationship between investment and output, is presented to explain, in a general way, the acceleration in the rate of growth of the Iraqi economy. To illustrate the point, consider a linearly homogeneous production function. Specifically, consider a Cobb-Douglas production function with the elasticity of output with respect to capital and the elasticity of output with respect to labor adding up to one. This kind of production function suits the purposes of this analysis because "many real-world production functions (at least for economies as a whole) are homogeneous of degree 1."[1]

This is not the place to discuss production theory. (The interested reader may consult any standard textbook on price theory.) Suffice it to point out here that this theory states that an increase in inputs should, ceteris paribus, result in an increase in output—that is, investment (capital accumulation) is a source of economic growth.

Now, it was shown in the preceding chapter that capital formation in Iraq has been growing at an accelerated rate. Specifically, total real GDFC rose from ID 269 million in 1973 to ID 489 million in 1974, ID 690 million in 1975, and ID 873 million in 1976. As a percentage of real GDP, real GDFC rose from 19.8 percent in 1973 to 42 percent in 1976. According to production theory, and barring capacity and underutilization and other obstacles, the observed acceleration in investment should result in an acceleration in output.

This, of course, is a general statement, and in the absence of more information as to the nature of the production function in Iraq, it is not possible to determine the exact impact of the rise in investment on the growth in output. However, let it be assumed for illustrative purposes that the production function of the Iraqi economy has constant returns to scale. In other words, in order for Iraq to reap full benefit from its capital investment, it should increase the supply of other factors, particularly labor, proportionately. At first, this appears to be an impossible task. However, recognizing that Iraq is a developing country, it is in a position to increase the supply of labor. Most developing countries, Iraq included, suffer from unemployment.[2] That is to say, the supply of labor can be increased in the short run through reduction and/or complete elimination of unemployment. As indicated earlier, Iraq also imported skilled labor in order to augment domestic supply. Therefore, a simultaneous increase in capital and decrease in unemployment should lead to accelerated growth in Iraq.

Alternatively, it is possible to rely on yet another feature of linear homogeneity that holds true in the short run. Under the assumption of a homogeneous production function of the first degree, it is possible to distinguish, in the short run, three stages of production. The first stage is characterized by increasing average product and, therefore, by increasing total product. This is so because the efficiency of the variable factor (capital in the case of Iraq at the present time) is increasing. The increase in total product also means that at this stage the efficiency of the fixed factor (labor) also is increasing.

In the second stage, the average product begins to decrease, owing to the decrease in the marginal product of the variable factor. However, total product continues to increase because the marginal product of the variable product, though falling, is still positive, and because the efficiency of the fixed factor is still increasing. The third and last stage of production is characterized by a negative marginal product of the variable factor, a declining average product, and a declining total product.[3]

It is not difficult to imagine that Iraq is now in the first stage of production and that capital deepening should improve the efficiency of both capital and labor. If this is accepted, then accelerated growth would be the only possible outcome, everything else being the same, at least in the short run.

Finally, recourse can be made to the neoclassical growth theory that views labor and capital as substitutes for each other. "If the rate of growth of the labor force is exogenously determined by demographic factors, capital may be treated as 'putty' which can then be shaped to absorb any size of labor force."[4] On this account, higher rates of output could be expected in Iraq, even though the labor force remains stable or does not increase at the same rate of increase in capital stock. The capital deepening that is going on in the country at the present time would serve to raise the productivity—that is, improve the efficiency of labor.

Post-Keynesian economists regard technical progress as an independent source of economic growth. Furthermore, it has been argued since 1960 that investment is a carrier of technological progress.[5] This certainly holds true in the case of Iraq. Development projects are implemented with the aid of foreign contractors, who are urged to bring into the country the most advanced technology at their disposal. In other words, new investment in Iraq does not only serve to deepen the stock of capital but also serves as a carrier of technology. Since technological progress is an independent source of economic growth, modern machines and equipment and advanced methods of production brought into Iraq with every new wave of investment should result in an increase in output.

To sum up, then, no matter how it is viewed, the expansion in the absorptive capacity of the economy—the rise in investment—points to an era of accelerated economic growth. It will be shown in the following subsection that recent economic developments in Iraq are in conformity with this theoretical discussion.

It is worth pointing out at this juncture that capital and advanced technology can replace certain types of labor as well as substitute for otherwise required increments of labor. For example, in some instances (for example, textiles, clerical work) it is quite possible to raise production substantially by increasing capital and applying modern techniques while reducing the amount of labor applied. The implication here, of course, is that unskilled labor may increasingly be in less demand and skilled labor even more in demand. The very sharp rise in the wages of skilled labor that is going on now in Iraq is a manifestation of this structural change. This explains, all the more, the desire of the Iraqi government to upgrade the labor force, since the income distributional effects of such a revised production function are a matter of serious concern.

Growth Rates Before and After the Rise in Oil Prices

From Table D.3, it can be ascertained that real gross domestic product rose from ID 1,360.0 million in 1973 to ID 1,499.2 million in 1974, ID 1,774.5 million in 1975, and ID 1,994.7 million in 1976. With the exception of agriculture and mining and quarrying (oil extraction), all other sectors registered positive and higher-than-normal rates of growth in 1974. In 1975 agriculture was the only sector to experience a real decline in output; and in 1976 the services sectors experienced a substantial (-23.1 percent) negative rate of growth, due to a deliberate action on the part of the government to restrain its current and development expenditures in order to contain inflationary pressures and rationalize investment.

Actual real rates of growth during 1974, 1975, and 1976 are presented in Table 7.1 together with historical growth rates over the 1954-73 and 1964-73 periods. In 1974 gross domestic product increased by 10.2 percent as against 6.0 percent over the 1954-73 period and 4.7 percent over the 1964-73 period. However, this is not indicative of the true performance of the economy. The overall rate of growth was held down by the decline in the oil sector (-3.1 percent). It is this fact that explains the 16.4 percent rate of growth in the nonoil GDP. Furthermore, since the growth of agricultural output was negative (-8.0 percent), the nonoil, nonagriculture GDP grew by about 22.4 percent in 1974—that is, more than three times the rate of growth over the 1964-73 period.

In 1975 the overall rate of growth accelerated to 18.4 percent in spite of the worldwide depression and the negative rate of growth of the agricultural sector. The decline in agricultural output perhaps was caused by adverse climatic conditions and most probably by movement of labor out of farming into other sectors where wages rose sharply.

TABLE 7.1

Aggregate and Sectoral Real Rates of Growth of the Iraqi Economy
Before and After the Rise in Oil Prices
(in percent)

	1954-73	1964-73	1974	1975	1976
GDP	6.0	4.7	10.2	18.4	12.4
Nonoil GDP	6.8	5.2	16.4	19.2	14.9
Nonoil, nonagricultural GDP	7.1	6.0	22.4	23.1	15.6
Agriculture	4.5	2.8	-8.0	-1.6	10.5
Mining and quarrying	4.7	3.8	-3.1	16.2	8.5
Industry	6.7	7.2	6.8	22.2	86.2
Distribution sectors	6.9	4.0	37.5	17.2	13.8
Services sectors	9.0	6.5	25.7	27.5	-23.1

Sources: Table 4.1 and Table D.3.

The rate of growth decelerated in 1976 to 12.4 percent. As explained earlier, this is attributable to the sharp decline in the rate of growth of the services sector. Other aggregate and sectoral rates indicated that the economy was growing vigorously. The industrial sectors as a group grew by 86.2 percent. The highest rate of growth occurred in the construction subsector (258.8 percent), indicating that the implementation of development projects was proceeding at a very fast rate. The manufacturing subsector registered a rate of growth of about 25.3 percent, also indicating that some of the new projects have come on stream. Finally, the electricity and water subsector grew by about 50 percent.

These results suggest that the Iraqi economy has entered an era of accelerated growth. Once again, the conclusion is tentative. Be that as it may, it should be pointed out that rates of growth during those three years are very impressive indeed. Their significance owes its origin to the fact that the rates of growth in 1974 were based on the actual performance of the economy during 1973 when GDP grew by 10.5 percent—that is, at a much higher rate than the rates observed over the 1954-73 and 1964-73 periods. Similarly, growth rates in 1975 and 1976 were based on actual performance in 1974 and 1975, respectively. As indicated earlier, real rates of growth during 1974 and 1975 were exceptionally high.

NECESSARY CONDITIONS TO SUSTAIN ACCELERATED GROWTH

An attempt was made in the preceding section to relate the high rates of economic growth in Iraq to the expansion in absorptive capacity. Therefore, it follows that the sustainability of accelerated growth would depend in the first instance on the country's ability to continue domestic investment at high levels. This in turn would depend on the removal of the constraints that limited its ability to invest in the past and on the maintenance of the terms of trade advantage it has been enjoying since 1974. This section is intended to deal with these two topics.

Development Policies

As can be seen in Table 7.2, the availability of oil revenues enabled Iraq to accelerate the implementation of its development projects and initiate new ones. Development expenditures rose from ID 196.9 million in 1973 to ID 509.0 million in 1974, ID 893.3 million in 1975, and ID 1,035.2 million in 1976. This is in accord with the stated policy of the government.

TABLE 7.2

Iraq: Actual Expenditures of the National Development Plan (1973-76)
(millions of Iraqi dinars)

Sector	1973	1974	1975	1976
Agriculture	37.8	78.0	99.9	190.8
Industry	66.4	184.1	290.2	515.8
Transport and communication	27.6	105.6	138.0	160.5
Building and services	36.5	90.6	101.0	118.1
Miscellaneous investment	28.6	51.6	264.2	50.0
Total	196.9	509.9	893.3	1,035.2

Source: Iraq, CSO, Annual Abstract of Statistics 1976.

It is widely believed that Iraq will achieve rapid development in the course of the coming years. After dropping in 1974 because of initial bottlenecks, the development implementation rate rose again, reaching a level of 85 percent in 1975. Further improvement is expected as a result of the extra efforts aimed at eliminating bottlenecks, particularly in the construction and transport and communication sectors, and increasing the supply and improving the quality of semiskilled and skilled labor.

Construction Material

The investment programs for 1974 and 1975 deliberately concentrated on projects requiring modest-to-moderate amounts of construction materials. At the same time particular emphasis was placed on accelerating the completion of brick, prefabricated construction materials, concrete blocks, and cement factories.

Projects included in the National Development Plan (1976-80) are designed to eliminate completely these bottlenecks and secure adequate supplies of construction materials. For example, existing plans call for increasing the amount of cement produced locally from 2.8 million tons in 1976 to 10 million tons in 1980.[6]

Transport and Communication Sector

Equally emphasized are projects designed to expand port, transport, and storage facilities. Most of these projects have been completed and the rest will be completed very shortly. A tangible result of this effort is the resolution of the problem of port congestion in Iraq.[7]

The inadequacy of the transportation system was demonstrated in 1974. The transportation crisis prompted the government to take urgent measures to deal with the situation. The determination of the government is demonstrated by the fact that real investment in this sector rose from ID 30.4 million in 1973 to ID 78.6 million in 1974 and ID 90.3 million in 1975.

The NDP (1976-80) aims at removing the transportation bottlenecks by expanding and improving the country's network. Specifically, the plan aims at expanding the capacity of the transportation sector by 10 percent annually. Port capacity will be increased from 4.75 million tons in 1975 to 14 million tons in 1980; and storage capacity will be expanded from 170,000 tons in 1975 to 370,000 tons in 1980.[8]

Manpower Planning

Shortages in skilled labor and middle managers are recognized to be the most serious limit on the absorptive capacity of the Iraqi

130 / OIL REVENUES & ACCELERATED GROWTH

economy. As indicated earlier, specific measures were taken to relieve the situation in the short run. In the medium and long run, the authorities intend to implement a plan designed to meet the country's manpower needs. The plan rests on reducing the proportion of the labor force engaged in agriculture, increasing the participation of women in the labor force, and pursuing educational programs especially designed to increase the supply of skilled workers and professionals.

This new educational policy lays particular stress on vocational and scientific training and research. For example, the educational plan envisages an increase in the number of students entering vocational schools upon graduation from the ninth grade from 7,745 during the academic year 1975/76 to 50,000 during the academic year 1980/81.[9] The number of graduates from higher technical institutes is anticipated to rise from 4,059 in 1976 to 6,779 in 1978.* Finally, the number of graduates from Iraqi universities is projected to increase from 11,845 in 1977 to 14,887 in 1980. Within these aggregate numbers, the ratio of university graduates specializing in humanities will decline from 32 percent of the total in 1977 to 13 percent in 1980.[10] This is, of course, in addition to the large number of Iraqi students abroad and in addition to the training institutes attached to most state organizations that are designed to serve their specific needs.

It should be pointed out that the government should have no difficulty reaching these goals. All schools and universities in Iraq are public, and admission is controlled so as to insure the desired balance between the various fields of learning.

Investment in Human Capital

Increasing the supply of skilled labor is an important objective. However, even more important is the objective of improving the quality of labor in general. This can be achieved only through improvements in the quality of life and higher standards of living. This is, of course, what development planning is all about.

The NDP (1976-80) is geared toward making rapid achievements in this field. In particular, health and educational services

---

*These institutes differ from vocational schools in that they only accept students who have completed secondary education (twelfth grade). They also differ from colleges in that they offer a concentrated program and award diplomas after three years instead of the customary four and sometimes five years of study required by all colleges in Iraq.

will receive special attention. In the area of health, the plan aims at increasing the number of medical doctors by 45 percent by 1980 so that there will be a doctor for every 2,000 persons. The number of hospital beds will be increased by 25 percent.[11] Most of the expansion will be in small towns and rural areas where, as everywhere in the country, the emphasis will be on preventive medicine.*

Other related programs include rural electrification, where it is planned to extend electric power to at least 65 percent of rural residents by 1980; provision of potable water to at least 65 percent of rural residents by 1980, as against 13 percent in 1975; and acceleration in the construction of sewage systems in major cities and the adoption of measures necessary for controlling and preventing pollution and protecting the environment.

In general, the government is intent upon increasing social services, particularly for low income classes. A measure of this policy is the fact that per capita expenditure on social welfare is projected to increase from ID 60 in 1976 to ID 79 in 1980 (in constant 1975 prices).[12] These expenditures, though classified as current, do indeed serve developmental purposes. As Theodore W. Schultz observed:

> Much of what we call consumption constitutes investment in human capital. Direct expenditures on education, health, and internal migration to take advantage of better job opportunities are clear examples. Earnings foregone by mature students attending school and by workers acquiring on-the-job training are equally clear examples. Yet nowhere do these enter into our national accounts. The use of leisure time to improve skills and knowledge is wide-spread and it too is unrecorded. In these and similar ways the quality of human effort can be greatly improved and its productivity enhanced. I shall contend that such investment in human capital accounts for most of the impressive rise in the real earnings per worker.[13]

Investment in human capital raises the earnings of workers because, in the long run, it leads to a reduction in the capital-output ratio.[14] Reduction in the capital-output ratio means an increase in the marginal productivity of labor (if there is no compensating increase in labor or decrease in output). If the economic

---

*Newly graduated doctors are required by law to serve at least one year in rural areas.

principle that wages should be equal to the marginal product of labor is to be operative, then real earnings per worker should increase as a result of investment in human capital.

Social welfare programs take time for their impact to be realized. However, there are other ways of improving the quality of life that can have an immediate impact. One way is to raise the purchasing power of the people. In this connection, the Iraqi government acted promptly following the rise in oil prices. The increase in oil prices went into effect on January 1, 1974; on February 7, 1974, the Revolutionary Command Council approved a series of decrees designed to immediately infuse oil revenues into the hands of the people. Decrees Nos. 95 through 98 of 1974 increased the minimum wage of unskilled laborers employed by the public sector, the cost of living allowances granted all classes of government employees, the salaries of some and the cost of living allowances of all military and police personnel, and the salaries of all retired government employees. Decrees Nos. 99 through 101 of 1974 reduced income taxes, real estate taxes, and indirect taxes; the prices of electricity, water, natural gas, kerosine, gasoline, and other goods and services provided by state-owned monopolies; and interest rates on housing loans provided by government-owned banks. Decree No. 103 of 1974 provided for the appointment to government posts of all unemployed college graduates no later than June 1, 1974.[15]

On April 3, 1974, the Revolutionary Command Council approved a law raising the upper limit of the salaries that government employees of all grades may receive.[16] Also during April, the minister of industry notified the General Federation of Labor that the minimum wage of unskilled laborers in the private sector had been raised.[17] The government hesitated at first to raise the minimum wage of this group because of fear that such an action would push the cost of locally produced goods and services up and intensify the anticipated inflationary pressure.[18]

The impact of these and similar measures on current consumption can be assessed from the significant rise in the government's current expenditures. Ordinary or current expenditures rose from ID 345.3 million in fiscal year 1972 to ID 777.3 million in fiscal year 1974 and were estimated to reach ID 2,616 million in fiscal year 1979.[19]

Increasing disposable income would be to no avail if goods and services were in short supply and if inflationary pressures (domestic or imported) were to be allowed to build up unabated. To meet the requirements of its ambitious investment programs and satisfy the rise in demand for consumption goods, the government adopted a liberal import policy. The minister of economics announced on February 13, 1974, that the government had allocated ID 1,134 million to finance the country's import program for fiscal year 1974, which was to commence on April 1, 1974, as against ID 350 million in fiscal year 1973.[20] The timing of the announcement was significant, as it came only five days after the Revolutionary Command Council approved Decrees Nos. 95 through 103, which were designed to raise the purchasing power of the people.

The liberal import policy caused the value of imported goods and services to increase from $1,650 million in 1973 to $5,874 million in 1975—that is, to increase by 256 percent in two years.[21] Although part of this increase is attributable to the rise in import prices, its bulk is due to growth in volume.

The increase in aggregate supply was augmented by the establishment of a Price Support and Stabilization Fund, "in order to accelerate the process of economic and social development, raise the standard of living of the people, stabilize the market, and enable the Central Pricing Organization to discharge its task."[22]

The program was successful in controlling inflation during the difficult years of 1974 and 1975. Consumer prices in Iraq rose by only 8.3 percent and 9.5 percent during those two years, respectively.[23] These results are very favorable in view of the fact that the worldwide rate of change in consumer prices was 15.2 percent in 1974 and 13.4 percent in 1975.[24] Iraq's performance appears to be indeed impressive when compared with the performance of OPEC and other developing countries. The average rise in consumer prices in OPEC countries was 15.5 percent in 1974 and 16.6 percent in 1975; it was 31.5 percent and 29.7 percent, respectively, in 1974 and 1975 in nonoil developing countries.[25]

The price stabilization program has its costs. For example, the budget of the Price Support and Stabilization Fund was increased from ID 135 million in 1977 to ID 250 million in 1978.[26] Another, and more serious aspect, is price distortions. The government should be aware that disparity between domestic and international prices could lead to shortages, smuggling, and black market operations and also could have an adverse effect on the allocation of resources.*

In defense of this policy, it should be pointed out that there are few countries in the world that do not have price support programs

---

*Iraq does not rely on free market forces to achieve an optimum allocation of resources. Resources are allocated directly through central planning, and, therefore, there is little risk that extensive price controls would cause a misallocation of resources.

of one kind or another. A case in point is the price support program for agricultural products in the United States. More important, the government policy should be defended on the grounds that it is designed to protect low income groups from sharp fluctuations in the international prices of foodstuffs and other basic commodities and protect the newly acquired purchasing power of the people. These objectives should prove rewarding in the long run. In addition to equity considerations, there is an economic rationale for raising current consumption. As pointed out by Gunnar Myrdal, current consumption does serve developmental objectives:

> First, large masses of people in underdeveloped countries suffer from malnutrition and other serious deficiencies in their standards of living, in particular lack of elementary health and educational facilities and extremely bad housing conditions and sanitation. These impair preparedness and ability to work and to work intensively; production is therefore held down. This implies that measures to raise income levels for the masses of people could raise productivity. In the opposite direction, the forced savings on the part of these masses, brought about by inflation and the usually high regressive taxation in underdeveloped countries, may make increased physical investment possible. But at the same time it holds down or can even decrease labor input and labor productivity.[27]

Summary

The Iraqi government is determined to accelerate the implementation of development projects. It was shown at the beginning of this section that the government has achieved satisfactory progress. Furthermore, it should be pointed out that expenditures under the National Development Plan do not include other investment expenditures such as those classified as current and included in the ordinary budget or those undertaken by autonomous government agencies[28] or investment expenditures undertaken by the private sector.

It also was shown that the authorities are paying particular attention to solving bottlenecks in the construction and the transport and communications sectors and deepening investment in human capital. These policies have already proved their worth, and in the long run they will result in easing the absorptive capacity limits and sustaining the accelerated growth the country has been enjoying in recent years.

Oil Revenues

The second necessary condition for continued accelerated growth in Iraq is the availability of oil revenues on the same order of magnitude that materialized in recent years. The level of oil revenues is a function of both exports and prices.

On the pricing front, it will be recalled that beginning in October 1973 pricing policies have been decided and enforced by OPEC. As a founding member of this organization, Iraq is committed to abiding by its decisions. The official policy of OPEC is the maintenance of oil prices in real terms. The periodic adjustments of these prices during recent years have been in keeping with this objective. It is, of course, extremely difficult to speculate about future developments. Factors that are likely to influence the price of oil in the coming years will be discussed briefly in the following chapter. Suffice it to state here that virtually all knowledgeable sources agree that the real oil prices are unlikely to decline. To this extent, Iraq as well as other exporters of oil do not face any serious deterioration in their terms of trade.

With respect to exports, it was pointed out in Chapter 2 that Iraq exports about 95 percent of its oil production. During 1978, Iraq's production amounted to 2.6 million barrels per day.[29] And, according to the minister of oil, production level was in the neighborhood of three million barrels per day during the first quarter of 1979.[30] The increase was part of OPEC's effort to compensate for the virtual shutdown of oil fields in neighboring Iran.

Production is a function of reserves. Officially, Iraq's proved reserves are estimated to be 34 billion barrels. Other sources estimate the country's reserves in excess of 100 billion barrels.[31] The actual level of reserves notwithstanding, it is obviously clear from the country's plan to develop its oil resources that it is capable of producing between 3 and 4 million barrels per day during the next ten or fifteen years. Developments beyond this time horizon would depend, then, among other things, on the actual level of proved reserves.

Exports, however, depend not only on the level of production capacity but, more important, on the level of demand for oil in consuming countries. Such demand, in turn, is linked to the rate of economic growth in these countries, the availability of oil from sources other than OPEC, and the rate at which alternative sources of energy are substituted for oil. These topics are very broad and require detailed analysis, which is beyond the scope of this book. Nevertheless, they will be dealt with briefly in the following chapter. As far as Iraq is concerned, it is reasonable to conclude that, given its substantial proved and possible reserves and barring any unforeseen developments, it can maintain or even expand its share of the growing world oil exports. And this is sufficient to guarantee the flow of oil revenues necessary to sustain domestic investments at a high level.

OIL AND PROSPECTS FOR ACCELERATED GROWTH

Oil revenues admirably served Iraq as the principal source of development finance. The sharp increase in oil prices resulted not only in an increase in export receipts but also in a significant increase in the real income of the country. This fact, more than any other, caused the expansion in the absorptive capacity of the Iraqi economy and paved the way for accelerated economic growth.

Its fiscal function notwithstanding, the oil sector remained, until recent times, a foreign enclave. Plans for the development of the oil sector and for its integration with the rest of the economy are in the implementation stage. Since it is not practical to present a complete description of the various projects, this section is confined to a general survey of the salient features of the overall plan and its implications for the rest of the Iraqi economy.

The most important objective has always been the establishment of national sovereignty over the oil resources of the country. The process began with Law No. 80 of 1961 and was completed with the nationalization of foreign oil companies between June 1972 and December 1975. Foreign oil companies currently operating in Iraq do so under service contracts with the Iraqi National Oil Company (INOC).

The country's oil reserves are believed to be much higher than officially acknowledged. Thus, the first task to be undertaken was the intensification of the exploration for and development of these oil reserves. Existing plans call for raising production capacity from 2.1 million barrels per day in 1974 to 6 million barrels per day in 1981.[32] Present capacity is about 4 million barrels per day.[33] Capacity refers not only to production but to transporting and loading capacity. Thus, exploration and development activities are carried out simultaneously with pipeline and port construction. Many projects have already been completed. Since the policy is to have the INOC take full control of production, marketing, and transportation operations inside and outside the country, the Iraqi Oil Tanker Company was established as a fully owned subsidiary of INOC. The nucleus of the Iraqi oil tanker fleet is already operational.

The policy aims at processing crude oil locally to the extent possible. It is in pursuit of this policy that the refinery capacity of the country was substantially expanded in recent years. Further expansion will be undertaken as soon as long-term contracts for refined products are concluded.

Another, and perhaps the most important, objective is the establishment of petrochemical industries. Given its oil and natural gas endowments, Iraq possesses a comparative advantage in this field (like other oil producing countries). In addition to diversifying Iraq's exports, petrochemical industries will produce essential inputs for domestic industries (for example, fertilizer for agriculture, synthetic rubber for rubber users, and synthetic fibers for the textile industry). Work on the largest petrochemical project commenced in 1977.

Now, what do these policies imply for the rest of the economy? Well, they imply the following:

Development of the country's oil wealth after years of neglect at the hands of foreign oil companies;

Production and export levels commensurate with the actual reserves of the country;

Higher levels of oil revenues to enable the country to accelerate its development program and simultaneously continue and expand its investment in human capital;

Expansion of the country's industrial base through industrial investment in general and the establishment of petrochemical industries in particular; and

Investment in the oil sector, which is indeed an important outlet for the surplus of domestic savings over domestic investment. (In fact, the bulk of investment expenditures classified as "miscellaneous" [Table 7.2] represents investment expenditures in the oil sector. It makes sense to develop the natural resources of the country, especially so since the alternative is a negative rate of return on idle deposits in foreign banks.)

CONCLUSION

Growth theory states that investment and technical progress are sources of economic growth. It further states that investment is a carrier of technological progress. Developments in Iraq since 1974 confirm these economic principles. The expansion in the absorptive capacity did not only deepen the capital stock but served as a transmitter of modern technology. The economy responded with higher-than-average rates of growth.

Since it was recognized that bottlenecks existed, particularly in the construction and the transport and communications sectors, extraordinary measures were taken to resolve these problems. This leaves the country with its most serious bottleneck, namely, shortages in skilled labor and middle managers. Manpower planning under the direct control of the country's central planning agency is being implemented. For the long term, investment in human capital commands top priority.

The oil situation suggests that Iraq, as a member of OPEC, is likely to maintain both its oil exports and its terms of trade advantage. This expectation together with the efforts mentioned above at removing bottlenecks should make it possible for the country to sustain the recently observed high rates of economic growth. The development and integration of the oil sector would play a decisive factor in this endeavor. Finally, as the modern sector (oil included) expands, its capacity to stimulate skills and abilities will also expand. That is to say, the country's ability to invest will improve in the future and so will its chances of achieving accelerated growth.

NOTES

1. Richard A. Bilas, Microeconomic Theory: A Graphical Analysis (New York: McGraw-Hill, 1967), p. 114.
2. Unemployment in Iraq amounted to 6.8 percent of the labor force in 1973. See Iraq, CSO, Annual Abstract of Statistics 1973, p. 358.
3. Bilas, Microeconomic Theory, p. 118.
4. Joseph Aschheim and Ching-Yao Hsieh, Macroeconomics: Income and Monetary Theory (Columbus, Ohio: Charles E. Merrill, 1969), p. 127.
5. Ibid., pp. 124-30.
6. Republic of Iraq, Law No. 89 of 1977. (The National Development Plan [1976-80]).
7. Middle East Economic Survey, February 14, 1977, p. 8.
8. Republic of Iraq, Law No. 89 of 1977.
9. Ibid.
10. Ibid.
11. Ibid.
12. Ibid.
13. Theodore W. Schultz, "Investment in Human Capital," in The Goal of Economic Growth, ed. Edmund S. Phelps (New York: W. W. Norton, 1962), p. 106. Quoted in Aschheim and Hsieh, Macroeconomics, p. 123.
14. Ibid. The reference here is to physical capital.

15. For the texts of these decrees see Ath-Thawra, February 9, 1974.
16. Ibid., April 7, 1974.
17. Al-Jumhuriyah, April 29, 1974.
18. These reservations were voiced by the chairman of the Economic Affairs Council in a public meeting held in Baghdad on March 3, 1974. See Ath-Thawra, March 4, 1974. Subsequent developments proved that the government did not have to set a minimum wage for the private sector. Information gathered by the writer shows that as demand for labor intensified during 1974 and 1975 unskilled laborers' money wages reached a level of ID 1.5 to ID 2.0, three to four times the minimum wage decreed by the government.
19. Middle East Economic Survey, April 16, 1979.
20. Ath-Thawra, February 14, 1974.
21. IMF, International Financial Statistics, February 1978, p. 171.
22. Ath-Thawra, August 7, 1974.
23. IMF, International Financial Statistics, February 1978, p. 29.
24. Ibid.
25. Ibid.
26. Ath-Thawra, February 7, 1978.
27. Gunnar Myrdal, "Growth and Social Justice," World Development 1 (1973): 119.
28. Expenditures of autonomous government agencies are budgeted to reach ID 3,569.3 million in fiscal year 1979. See Middle East Economic Survey, April 16, 1979.
29. Oil and Gas Journal, February 26, 1979, p. 166.
30. Middle East Economic Survey, April 9, 1979.
31. Washington Post, August 7, 1978, p. A 10.
32. Petroleum Intelligence Weekly, June 10, 1975, p. 5. According to the minister of oil, production capacity was 4 million barrels per day during the first quarter of 1979. See Middle East Economic Survey, April 9, 1979.
33. Ibid.

8

OPEC:

PERFORMANCE AND PROSPECTS

W. W. Rostow theorized that "take-off" into sustained economic growth could be triggered by a shock—for instance, a revolution, a war, or a sudden and sharp increase in export earnings. His theory seems to be relevant to the recent experience of the oil exporting countries. Their export earnings have increased sharply, and, therefore, they ought to achieve self-sustaining growth.

Yet, such an outcome is not assured. Experience of other export economies suggests that it is entirely possible for export earnings to be spent on luxury imports and other unproductive outlays. Thus, a basic requirement of take-off is that oil revenues be invested domestically. Although it is understood that this must be done, the oil exporting countries, like other developing countries, face a constraint known as absorptive capacity.

The principal hypothesis investigated in this study was whether the sharp rise in oil prices resulted in an expansion of the absorptive capacity of the Iraqi economy. The main conclusions that emerged are the following:

The rise in oil prices had a positive impact on the absorptive capacity of the Iraqi economy. Absorptive capacity was doubled in less than two years.

The impact is mainly due to the positive income effect of the change in Iraq's terms of trade. It is also due to the fact that oil revenues made it possible for Iraq to augment its domestic factors of production through importation.

As might be expected, the expansion in absorptive capacity was accompanied by higher-than-normal rates of economic growth.

Future economic developments will depend on the availability of oil revenues and the maintenance of the terms of trade advantage.

At this stage it seems appropriate to survey, in a brief and general way, the overall performance of the oil exporting countries and assess their future prospects. The rationale for this concluding chapter follows from the fact that recent economic developments in Iraq were a reflection of developments in the oil sector, and those were the result of collective action taken by OPEC. More important is the fact that in the future Iraq's oil policy will not be independent from the overall policy stance of OPEC. Furthermore, this general survey will show whether economic developments in Iraq were part of a general trend spreading across the economics of the oil exporting countries.

THE OPEC EXPERIENCE

Developments in Iraq following the rise in oil prices were typical of developments in other oil exporting countries. Initially, most of them adopted expansionary fiscal and monetary policies in an effort to accelerate the pace of their economic development. In addition to emphasis on social services, a large share of government spending was directed to infrastructure in order to insure effective absorption of oil revenues.

Stepped-up development programs generated higher incomes and higher levels of domestic demand. Imbalances between aggregate demand and supply emerged soon thereafter and caused strong inflationary pressures. Consumer prices in the major oil exporting countries* rose by 17 percent in 1974 and 19 percent in 1975.[1] To satisfy the surge in demand, these countries expanded their import programs very rapidly. The total value of their merchandise imports rose by 77 percent in 1974 and by 57 percent in 1975. However, the sharp increase in the prices of OPEC imports (the unit value of OPEC's merchandise imports rose by 29 percent in 1974 and 10 percent in 1975) added to the upward pressure on domestic prices.

Faced with these difficulties, most members of OPEC found it necessary to change their policy stance. Late in 1975 and during 1976 there was a general tendency to restrain government spending in order to narrow the gap between domestic demand and available supplies. This was especially true in countries with a relatively high absorptive capacity. High absorbers were not only faced with the problem of inflation but also with rapidly vanishing current

---

*The major oil exporting countries are those listed in Table 8.2.

surplus (the combined OPEC current account surplus declined from $68 billion in 1974 to $9 billion in 1978). Fiscal developments in the major oil exporting countries during this period were summarized by one writer as follows:

> The growth of government expenditure in the major oil exporting countries has, by any standards, been spectacular; in some respects, it has been more spectacular than the growth in government revenues. Aggregate government expenditures increased by 49 per cent in 1973, doubled in 1974, and increased by a further 50 per cent the following year. Thereafter, the rate of growth in spending declined to 28 per cent in 1976 and 24 per cent in 1977. Budget estimates implied a further deceleration (to 16 per cent) in 1978. The pattern is similar for both high- and low- absorption countries, except for 1973 and 1976, when the growth rates were considerably higher in the low-absorption countries.[2]

Despite the problems of inflation and physical and human bottlenecks, the major oil exporting countries have achieved respectable rates of economic growth. Their combined real GDP grew, on average, by 5.8 percent during the 1974-78 period. In comparison, the average rate of growth during the same period was 4.9 percent in nonoil developing countries and 2.4 percent in industrial countries (Table 8.1).

TABLE 8.1

Changes in Real Output, 1974-78
(percentage changes in real GDP or GNP)

	1974	1975	1976	1977	1978
Industrial countries	0.1	-0.9	5.3	3.8	3.7
Developing countries					
Nonoil developing countries	5.3	4.0	4.9	5.1	5.2
Major oil exporting countries					
Oil sector	-1.0	-11.1	13.7	1.8	-4.5
Nonoil sectors	12.3	12.4	11.9	9.8	7.7
Total GDP	8.0	-0.3	12.8	6.2	2.5

Sources: Data from unpublished IMF material.

These global indicators do not reveal the true performance of the oil exporting countries. The average rate of growth is understated as a result of the stagnation in the oil sector. For example, real output in this sector declined by 1.0 percent in 1974, 11.1 percent in 1975, and 4.5 percent in 1978. The average rate of growth of the oil sector during 1974-78 was negative (-0.2 percent).

A more accurate picture is obtained by excluding the oil sector. During the recession years 1974 and 1975, nonoil real GDP grew by 12.3 percent and 12.4 percent, respectively. High rates of growth were maintained in 1976 and 1977. The noticeable decline in 1978 (to 7.7 percent) was partly due to developments in Iran where nonoil GDP declined by 1.2 percent.

These results bring out the blessings as well as the drawbacks of oil. Oil revenues have made it possible for these countries to undertake ambitious development programs and achieve comparatively higher rates of economic growth. The drawbacks, however, follow from the prominence of the oil sector. Performance of other sectors depends on the performance of the oil sector. The ups and downs of oil are simultaneously reflected throughout the economy.

Although OPEC has control over oil prices, it has no such control over the level of demand for oil. To avoid confusion, the reader is reminded that in the short and perhaps the medium run demand for oil is price inelastic. The vitality of the oil sector is linked to the vitality of industrial economics. Thus, when in 1975 the rate of growth of industrial countries declined by less than 1 percent, the rate of growth of the oil sector declined by more than 11 percent. The picture was reversed in 1976. The 5.3 percent rate of growth in industrial countries was matched by the 13.7 percent rate of growth in the oil sector (Table 8.1).

In general, the economic health of developing countries depends on the economic health of the industrial countries. This dependency is more pronounced in the case of oil exporting countries. The evidence is presented in Table 8.1. During 1974-78, nonoil developing countries maintained a stable, albeit below normal, rate of economic growth. On the other hand, the combined rate of growth in the major oil exporting countries fluctuated sharply as a consequence of economic fluctuations in industrial countries. Nevertheless, it should be recalled that the average rate of economic growth in the oil exporting countries over the 1974-78 period was higher than in nonoil developing countries and that it was attributable to the terms of trade income effect. In the absence of this terms of trade advantage, the impact of simultaneous recession and inflation in the industrial countries would have been felt with much greater severity in the oil exporting countries.

The degree of fluctuation differed from one oil exporting country to another. In countries where the relative size of the oil sector is large vis-à-vis the rest of the economy, fluctuations were more pronounced. This is evident from data presented in Table 8.2. For example, the overall rate of growth fluctuated widely in Saudi Arabia, while it remained more or less stable in Indonesia. It should also be noted that with the exception of the United Arab Emirates (UAE), Iraq, and Oman, all other major producers suffered a decline in their oil production in 1975. This was, as stated earlier, attributable to the recession in industrial countries. In most of these countries, the reduction ranged from 16 to 21 percent. It should also be noted that the increased output in the UAE, Iraq, and Oman was too small to compensate for the total reduction, as the combined total output of these three countries was less than half the total output of Saudi Arabia. Finally, the decline in Kuwait and Libya was not occasioned by slack in demand alone but was a consequence of a deliberate policy of conservation.

Saudi Arabia was perhaps happy to see a decline in its oil production, since it is agreed by all that this country's production far exceeds what is necessary to meet its financial requirements. Yet, it is Saudi Arabia and other large producers (namely, Kuwait, Libya, Qatar, and the United Arab Emirates) that are under pressure to increase their oil production in the name of international cooperation. The dilemma faced by this group is a serious one. They have small populations and their domestic investment opportunities are rather limited. They are, in fact, under obligation to conserve their most valuable natural resource in order to safeguard the interests of future generations.

Excessive production will subject them to another type of pressure. They will be urged to expand their domestic programs in order to reduce the size of their current account surplus. Again, this will be done in the name of international cooperation and the enhancement of the international adjustment process. However, to do so they would have to rely, as they have done so far, on a large contingent of expatriate labor. Ambitious development programs that are beyond sustainable levels and reliance on large numbers of foreign labor could have cultural, social, and political consequences that might not be in the best interest of these countries in the long run. This is one of the drawbacks of oil that is not fully recognized or understood.

Table 8.2 illustrates the differences that exist between the individual members of OPEC. In countries where the size of the oil sector is relatively large, the other sectors have achieved very high rates of growth, reflecting the substantial benefits that accrued to the rest of the economy as a result of the sharp increase in oil

TABLE 8.2

Major Oil Exporting Countries: Changes in Real GDP
(percent)

	1974	1975	1976	1977	1978
Algeria					
Oil sector	-8.6	-1.9	10.2	6.0	8.5
Other sectors	8.1	11.5	9.4	9.5	8.0
Total GDP	1.1	6.4	9.7	8.3	8.2
Indonesia					
Oil sector	2.9	-4.9	15.0	11.9	-3.0
Other sectors	8.3	6.2	6.0	6.4	7.3
Total GDP	7.6	5.0	6.9	7.0	6.1
Iran					
Oil sector	3.4	-12.2	9.5	-4.0	-8.1
Other sectors	14.2	15.5	13.4	8.7	3.0
Total GDP	8.8	2.2	11.8	3.6	-1.2
Iraq					
Oil sector	-3.1	16.2	8.5	2.5	8.2
Other sectors	16.4	19.2	14.9	10.0	12.0
Total GDP	10.2	18.4	12.4	7.9	11.0
Kuwait					
Oil sector	-15.7	-18.2	3.2	-8.5	6.0
Other sectors	7.0	12.0	14.5	7.2	10.0
Total GDP	-11.4	-11.4	6.4	-3.7	7.4
Libya					
Oil sector	-30.0	-2.8	30.6	7.5	-4.0
Other sectors	23.3	14.6	13.0	10.5	8.0
Total GDP	-13.3	5.0	22.0	8.9	1.5
Nigeria					
Oil sector	9.0	-21.0	16.3	—	-9.0
Other sectors	4.6	13.6	16.2	—	8.7
Total GDP	6.3	-0.6	16.2	8.7	3.4
Oman					
Oil sector	-1.0	17.8	7.3	-7.2	-7.7
Other sectors	91.2	27.3	13.7	-9.3	—
Total GDP	39.4	23.5	11.3	8.5	-2.8
Qatar					
Oil sector	-10.4	-15.9	16.0	-11.0	9.1
Other sectors	13.1	15.0	10.0	5.0	—
Total GDP	-4.0	-6.0	13.6	-4.9	5.3
Saudi Arabia					
Oil sector	10.8	-16.1	20.8	7.0	-9.8
Other sectors	16.4	19.1	21.9	19.5	15.9
Total GDP	11.5	-10.1	21.0	9.4	-4.4
United Arab Emirates					
Oil sector	8.2	1.3	14.4	2.8	-8.4
Other sectors	38.7	21.6	14.9	11.0	3.0
Total GDP	14.5	6.3	14.5	5.1	-5.0
Venezuela					
Oil sector	-12.8	-21.9	1.2	-3.0	-3.3
Other sectors	10.2	10.2	8.6	8.0	8.0
Total GDP	5.9	5.2	7.8	6.8	6.9

Source: Data from unpublished IMF material.

prices. However, in these countries fluctuations in the rates of growth were the sharpest. On the other hand, the rates of growth in the nonoil sectors in countries where the oil sector is not so large have not been so impressive, though they have been following a more stable course.

The size of the oil sector is, of course, only one element in a wider picture. There are other, more complex factors that account for the differences in the performance of individual countries. Analysis of these factors is beyond the scope of this book. Suffice it to state here that the major oil exporting countries

> were faced with difficult and novel issues and choices regarding the feasible speed of development of their non-oil sectors, the appropriate roles of the public and private sectors, tolerable rates of inflation, and acceptable standards of income distribution. Decisions on these issues differed somewhat from country to country, depending, inter alia, on the size of each country's population and oil reserves, its stage of development, and the domestic resources other than oil available to it. While varying emphases were, of course, observed, all countries aimed for a rapid increase in the rate of growth of the non-oil sector.[3]

FUTURE PROSPECTS

As a group, oil exporting countries have achieved a relatively higher rate of economic growth than the one achieved by nonoil developing countries. Especially encouraging has been the vigorous growth of the nonoil sectors in most of these countries. Their economic performance is certainly attributable to the sharp increase in oil revenues and the terms of trade income effect advantage they have enjoyed since 1973. It, therefore, follows that their future prospects would depend, among other things, on the availability of oil revenues and the preservation of the terms of trade advantage.

Oil revenue is a function of both oil exports and prices. The maintenance of the terms of trade advantage means that real oil prices will not decline and that the real prices of OPEC's imports will not increase in the future. Alternatively, price changes in OPEC's exports and imports could offset each other. These are wide and complicated issues that cannot be dealt with adequately here. What follows is a brief and speculative discussion (since the future is highly unpredictable) of possible future developments with regard to demand for OPEC's oil and oil prices.

Demand for OPEC Oil

Under present circumstances, the determinants of the volume of oil exports from OPEC are the growth in demand for energy in consuming countries, the availability of oil from non-OPEC sources, and the rate at which alternative sources of energy are substituted for oil.

Growth in demand for energy and, therefore, in demand for oil depends very much on the rate of economic growth in consuming countries. The relationship between economic activity and demand for oil was demonstrated during the recession years of 1974 and 1975. During each of those two years, world consumption of oil declined by more than 3 percent. Consequently, the exports of the major oil exporting countries declined from 29.5 million barrels per day (mbd) in 1973 to 29.1 mbd in 1974 and 25.6 mbd in 1975. Responding to the partial economic recovery, these exports rose to 29.1 mbd in 1976 and 29.2 mbd in 1977.

Future developments will depend to a great extent on the level of economic activity in industrial countries. In this connection, a recent World Bank report projected the average annual rate of growth of GNP in the industrial countries at 4.8 percent during 1978-80 and at 4.5 percent during the 1978-85 period.[4] The International Monetary Fund reported that the actual rates of growth of GNP in the industrial countries were 5.4 percent during 1976 and 3.7 percent during 1977.[5] It anticipates a rate of growth of 4 percent during 1978 and believes that a rate of growth of 4.75 percent over the two-year 1979-80 period is feasible.[6] Both institutions appear to be in agreement that a recurrence of the deep recession of the mid-1970s is unlikely.

As regards the availability of oil from non-OPEC areas, the two most promising sources are Alaska and the North Sea. There has been much delay in utilizing these reserves. Alaskan production commenced in 1977 at the rate of 0.3 mbd and was expected to rise to 1.1 mbd in 1978. Norway's production began in 1975 at the rate of 0.2 mbd and rose to about 0.35 mbd during 1978. Finally, the United Kingdom commenced production in 1976 at the rate of 0.2 mbd. Its production reached a level of 1.1 mbd in 1978.[7] Other new sources include Egypt and Mexico, where production in 1978 amounted to 0.5 mbd and 1.2 mbd, respectively.[8]

Against these advances in oil production outside OPEC, one must take into account the decline in Canadian production from 2.1 mbd in 1973 to 1.3 mbd in 1978 and the decline in U.S. production from 11.0 mbd in 1973 to 8.7 mbd in 1978.[9] Thus, excluding the communist countries, world oil production outside OPEC has not changed much; it amounted in 1978 to 17.1 mbd, the same as it was in 1973.[10]

The third and last determinant of demand for OPEC oil is the rate at which alternative sources of energy are substituted for oil. It is not altogether certain that these alternative sources will become available before the mid-1980s. In addition to technical difficulties and constraints imposed by environmental considerations, the whole question of supply programming for new energy sources in industrial countries is likely to be hindered by three sets of economic difficulties.

> First, the inter-country differences in energy-resource endowments raise the problem of cost-sharing in the development of new energy sources. . . . it is obvious that several forms of international cooperation are needed at R & D, investment, production and distribution stages. Secondly, at the national level in energy-rich countries, the development of indigenous energy sources inevitably entails the thorny problems associated with the collection of economic rents that will potentially accrue to holders of energy assets while also taking into account the need to provide incentives to energy-related investments. . . . Thus, the acceleration of indigenous energy production in energy-rich areas requires politically difficult adjustments in fiscal and income policies. The third set of difficulties surrounding energy resource developments is related to the size and structure of future capital formation activities. In view of the currently observed shortfall of capital formation in the materials producing sectors of potentially energy-rich countries, such as the U.S.A., it will be the important task of economic development policy to accommodate energy-related capital requirements in the national investment program.[11]

These difficulties underscore the possibility that it might be a long time before the development of alternative energy sources is tackled in a major way. This, of course, is regrettable. New sources of energy would be of great benefit not only to consumers but especially to members of OPEC who are concerned about the rapid rate of depletion of their nonrenewable resource.

To conclude, then, based on prospects for industrial countries, the availability of oil from non-OPEC sources, and the rate at which alternative sources of energy might be developed, demand for OPEC's oil is likely to increase between now and 1985. The World Bank expects OPEC net oil exports to amount to 31.8 mbd in 1980 and to 33.9 mbd in 1985.[12]

Oil Prices

Beginning in October 1973, the pricing policy has been completely in the hands of OPEC. As stated by former Secretary General M. O. Feyide, OPEC prefers to base pricing decisions on these major considerations:

> The sovereign right of member countries to decide the price of their crude oil.
>
> The need to supply consuming countries with adequate amounts of crude at "fair and reasonable prices," taking into account the intrinsic value of petroleum as an exhaustible asset and the cost of alternative energy.
>
> Maintenance of the purchasing power of petroleum revenue.[13]

The secretary general stated that in setting future prices OPEC will also consider "the price of alternative energy, the rate and size of new discoveries of oil, energy conservation measures, and governmental policies in developed countries with particular reference to inflation, currency variations, and rates of economic growth."[14]

Given OPEC's official position, it is not possible to anticipate a price reduction. On the other hand, it is possible to argue that OPEC's control over prices is not as firm. There are two arguments that can be advanced in support of this hypothesis. First, OPEC will soon be facing competition based on discovery and developments of non-OPEC oil and the availability of alternative sources of energy. Second, OPEC might crumble as individual members reduce prices in order to increase their market share and maintain required minimum incomes.

Neither of these two developments is likely to materialize. As regards competition, it must be pointed out that owners of new oil want to sell at OPEC prices. Addressing a conference held in London during November 1975, Lord Balogh, U.K. minister of state for energy, stated that the British government will be "price takers and not price makers" for its North Sea participation crude.[15] The same holds true for Mexico and other new producers.

The second factor that OPEC must take into account is the cost of alternative sources of energy. The cost of alternative energy is a controversial issue. Nevertheless, there is general agreement that it will be considerably higher than the present price of crude oil. Commenting on this subject a knowledgeable source observed that:

> Ultimately, we are going to have to decide what is a
> fair price, and I don't think there is any other defini-
> tion of "fair" than the cost of marginal energy. I don't
> mean nuclear energy; I mean the cost of synthetic,
> liquid hydrocarbons. I've used the figure $16 to $18;
> Continental Oil Company, you may have seen, has just
> come up with a study saying it's between $20 and $24 a
> barrel. In other words, a long way from the present
> price. That will probably be the ultimate price; the
> price we will have to live with, ultimately.[16]

Regarding the possibility of an intra-OPEC competition for market shares, it should be explained that OPEC enjoys a thick cushion between demand for its oil and its minimum production level. The minimum production level is that point below which OPEC as a whole becomes uncomfortable for lack of revenue; it is determined by the maximum amount of money members of OPEC can spend effectively for development and other programs and the volume of oil sales required to generate this amount of money. Assuming a $10 per barrel take and 1974 prices, Shell Oil Company economists, H. de Vries and C. A. Gabelein, estimated the minimum production level to be in the range of 12.9 to 15.5 mbd in 1975 and in the range of 15.9 to 22.5 mbd in 1980.[17]

Thus, comparing the level of demand with minimum OPEC production, it can be seen that the demand for OPEC oil exceeds minimum production by a significant amount. This fact precludes any possible price reduction in order to increase market share.

In addition, members of OPEC are not likely to resort to price cutting while they know they can increase their revenues by limiting production and raising prices:

> The projections for OPEC oil exports derived under
> alternative assumptions for world oil prices suggest
> that oil revenues to be generated from restricted vol-
> umes at higher prices are likely to be larger than those
> generated from higher volumes at lower prices in the
> remaining part of the current decade. Such prospects
> may continue to hold in the earlier part of the 1980s,
> until the stage when substantial amounts of primary
> energy from alternative sources become available in
> the OECD area.[18]

One such projection is that made by Hendrik S. Houthakker. Using the "Economic Model of World Oil Market"[19] developed by M. Kennedy and making two sets of assumptions about price and income elasticities, Houthakker concludes that (1) under "optimistic" elasticity assumptions, OPEC would maximize its oil revenue in 1980 by charging a tax of $7.50 per barrel and limiting output to 15.4 million barrels per day and (2) under "pessimistic" elasticity assumptions, OPEC would maximize its oil revenue in 1980 by charging a tax of about $20 per barrel and limiting output to about 12.1 million barrels per day.[20] Houthakker, however, thinks that OPEC will not be able to cut output so drastically because such a move will jeopardize its unity.

In brief, taking into account economic factors alone could lead to the inevitable conclusion that the price of crude oil will not decline anytime soon. "On balance the analysis of the oil market suggests that average oil prices could be maintained at about $11.50 a barrel (in 1975 US dollars) to the early 1980s."[21]

Yet, it is not appropriate to base future price developments on consideration of economic factors alone. Political considerations play a major role. They, more than any other factor, were responsible for OPEC's failure to raise prices during 1976. In a clear political gesture to President Carter who was about to assume office, Saudi Arabia prevailed upon the other members of OPEC to limit the increase in prices to 10 percent during 1977.[22] Of course, no one knows if Saudi Arabia will continue to play such a role. Much will depend on the ability of industrial countries to contain inflationary pressures and the development of alternative sources of energy.

To conclude, then, reductions in the demand for OPEC's oil or in the real prices of oil do not seem likely under present circumstances. And this means that the major oil exporting countries have an excellent opportunity to maintain and further the economic progress they have achieved during the recent past.

NOTES

1. Data quoted in this section were provided by the IMF.
2. David R. Morgan, "Fiscal Policy in Oil Exporting Countries, 1972-78," IMF Staff Papers 26 (1979): 65-66.
3. Ibid., p. 64.
4. Helen Hughes, et al., Prospects for Developing Countries, 1978-85 (Washington, D.C.: IBRD, 1977), p. 16.
5. IMF Survey, May 8, 1978, p. 137.
6. Ibid., p. 141.
7. Oil and Gas Journal, February 26, 1979, p. 166.
8. Ibid.
9. Ibid.
10. Ibid.

11. Merith Celasum and Frank Pinto, "Energy Prospects in OECD Countries and Possible Demand for OPEC Oil Exports to 1980," mimeographed, World Bank Staff Working Paper, No. 221 (Washington, D.C., September 1975), pp. 16-17.

12. Hughes, Prospects for Developing Countries, 1978-85, p. 19.

13. Oil and Gas Journal, December 1, 1975, p. 90.

14. Ibid.

15. Ibid.

16. Address to the Third International Conference sponsored by the University of Colorado's International Research Center for Energy and Economic Development by Mr. James E. Akins, former head of the State Department's Office of Fuels and Energy, Boulder, Colorado, October 18-19, 1976. Reprinted in Middle East Economic Survey, November 1, 1976, pp. 4-5.

17. Oil and Gas Journal, December 1, 1975, p. 88.

18. Celasum and Pinto, "Energy Prospects in OECD Countries," p. 14.

19. For a description of this model see M. Kennedy, "An Economic Model of the World Oil Market," Bell Journal of Economics and Management Science 5 (1974): 540-77.

20. Hendrik S. Houthakker, The World Price of Oil: A Medium Term Analysis (Washington, D.C.: American Enterprise Institute for Public Policy Research, 1976), pp. 17-24. The tax rates do not represent the price of oil. In order to determine the price, it is necessary to add production cost and company margin. Dollar figures refer to 1974 dollars.

21. Hughes, Prospects for Developing Countries, 1978-85, pp. 18-19.

22. Washington Post, May 6, 1979, p. A 18.

APPENDIX A
NOTES ON DATA

In the course of this study, data on Iraq's foreign trade, gross domestic product, gross domestic fixed capital formation, and development expenditures were assembled. This data bank was drawn upon for purposes of quantitative analysis particularly in Chapters 4, 5, 6, and 7. A note on the manner in which these data were gathered and organized is deemed to be appropriate.

Most of the data was gathered from official Iraqi sources, namely, the Central Bank, the CSO, and the Ministry of Planning. No particular explanation is called for in the case of data expressed in current prices. However, it is necessary to offer some comments on data expressed in constant 1969 prices. Data in Table 5.2, Table D.3, and Table D.5 are expressed in this manner.

TABLE D.3—GDP

Three sources were relied upon to construct the time series on Iraq's gross domestic product: Jawad Hashim, Hussein Omar, and Ali al-Munoufy, The Planning Experience; CSO, Annual Abstract of Statistics; and the Social and Economic Data Bank of the World Bank. The World Bank assembled data on Iraq's GDP in both current and constant 1969 prices for the 1960-75 period. It obtained data from the Central Statistical Organization, and they correspond to data published in the various issues of the Annual Abstract of Statistics. Using a series of implicit GDP deflators, the World Bank converted these data into constant 1969 prices. The 1976 issue of the Annual Abstract of Statistics contains data on Iraq's GDP expressed in constant 1969 prices—for the years 1973, 1974, and 1975. These data correspond to data obtained from the World Bank. Thus, it became possible to assemble with confidence data on Iraq's GDP for the 1960-75 period expressed in constant 1969 prices. Data on Iraq's GDP in 1976 in current prices were provided by CSO. They were converted into constant 1969 prices by the author.

It was not possible to gather data on Iraq's GDP for the 1953-59 period expressed in 1969 prices. However, The Planning Experience contains data on Iraq's GDP for the 1953-69 period expressed in constant 1964 prices. The 1964 implicit GDP deflator provided by the World Bank was used to deflate the figures for the 1953-59 period and express them in constant 1969 prices. The figures for some overlapping years also were deflated and found to be almost identical to the figures obtained from the World Bank. This result strengthened the confidence in data available in <u>The Planning Experience</u>.

TABLE 5.2—GDFC

In addition to the three sources cited above another source, namely, Jawad Hashim, <u>Fixed Capital Formation in Iraq 1957-1970</u>, was relied upon to assemble data on gross domestic fixed capital formation in both the public and private sectors for the 1953-73 period. Except for the years 1973-75, there are no published figures on Iraq's GDFC in both sectors and in constant 1969 prices. However, <u>Fixed Capital Formation</u> contains such data expressed in constant 1962 prices. Various issues of the <u>Annual Abstract of Statistics</u> also contain similar information for the 1964-75 period. For overlapping years, the figures in <u>Fixed Capital Formation</u> correspond to figures available in <u>Annual Abstract of Statistics</u>. The 1962 implicit GDP deflator provided by the World Bank was used to deflate figures for the 1957-72 period gathered from <u>Fixed Capital Formation</u> and <u>Annual Abstract of Statistics</u> and express them in constant 1969 prices.

It was not possible to obtain data on GDFC in both the public and private sectors expressed in constant prices of any year for the 1953-56 period. However, such data are available in current prices and they are to be found in Table 5.1. Also available are data on total GDFC expressed in 1969 prices for the years in question. These data are to be found in Table D.5. The relative shares of these two sectors in GDFC were calculated for the years 1953-56 from data in Table 5.1, and the percentages obtained were used to derive their respective shares expressed in constant 1969 prices from the totals available in Table D.5. The exercise was repeated for some later years and the results were found to be consistent.

TABLE D.5—GDFC BY ECONOMIC SECTOR

Data on gross domestic fixed capital formation by economic sector for the 1957-73 period expressed in constant 1969 prices were derived from the same sources and in exactly the same manner used to derive corresponding data in Table 5.2. Data on GDFC by economic sector for the years 1953-69 are available in <u>The Planning Experience</u> and are expressed in constant 1964 prices. The 1964 implicit GDP deflator obtained from the World Bank was used to deflate the figures for the 1953-56 period and express them in constant prices. The exercise was repeated for some overlapping years and the results were found to be consistent. Data for 1974 and 1975 were published in <u>Annual Abstract of Statistics 1976</u>.

OTHER REMARKS

Jawad Hashim is a former minister of planning in Iraq; his coauthors, Hussein Omar and Ali al-Munoufy also were attached to the Ministry of Planning. Since their works were published by this ministry, it can be assumed for all practical purposes that the data contained therein are official data.

The choice of 1969 as a base year was dictated by available data. Both the World Bank and the Central Statistical Organization chose 1969 as a base year. One can only speculate as to their reasons. The 1960-69 period was a period characterized by stable prices. The average inflation rate throughout this period was about 1 percent per annum. Prices began to rise at a much higher rate in 1970 and thereafter. For example, the inflation rate was about 7 percent in 1970. It should be pointed out that the oil prices began to rise in the last quarter of 1970. These considerations could have contributed markedly to the choice of 1969 as a base year.

APPENDIX B
ORDINARY LEAST SQUARES METHOD

The growth rates of the various sectors of the Iraqi economy presented in Table 4.1 and the growth rates of gross domestic fixed capital formation in Iraq presented in Table 5.3 were comuted according to the "ordinary least squares" methodology. The reasons justifying the use of this method were discussed in Chapter 5. The detailed results of the regression are presented in Table D.6 (economic rates of growth) and Table D.7 (growth rates of GDFC). Following is a brief description of the method used.

A subroutine that computes growth rate over a specified period by fitting a semilog least squares trend line to available data was used for this purpose. The functional form is as follows:

$$Y_t = Y_o (1 + R)^t \qquad (B.1)$$

where

t = time and
R = compound rate of growth per time period (year)

This rate of growth can be computed by the ordinary least squares technique by rewriting equation (B.1) in the following regression model:

$$Y_t e^\varepsilon = A (1+R)^t e^\varepsilon \doteq AB^t e^\varepsilon$$

or

$$\ln Y_t = \ln A + t \ln B + \varepsilon$$

where

$\ln Y$ = the dependent variable (in this case Y is gross domestic fixed capital formation in Iraq)

$\ln A$ = the intercept (constant) term
t = the independent variable
$\ln B$ = the ordinary least squares coefficient (which is the slope for the semilog trend line)
ε = the error term

Since

$$\ln B = \ln (1 + R)$$

then

$$R = EXP (B) - 1$$

Multiplying R by 100 gives the desired percentage rate of growth.

APPENDIX C
TERMS OF TRADE INCOME EFFECT

In Chapter 6, real income was defined as follows (equation [6.4]):

$$GDY_k = GDP_k + TTE$$

The derivation of this equation is presented in this appendix.* As explained in Chapter 6, in order to measure the terms of trade income effect, one needs only to calculate the difference between exports in the sense of "capacity to import" and exports measured in the constant price of the base period. Assume:

GDY = gross domestic income in current prices
GDP = gross domestic product in current prices
C = consumption in current prices
I = investment in current prices
E = value of exports measured in current prices
M = value of imports in current prices
P_e = price index for exports
P_m = price index for imports
k = subscript indicating base period prices
cm = subscript indicating capacity to import and
TTE = terms of trade income effect in base period prices

With the help of these notations it is possible to express exports in constant prices as:

$$E_k = E/P_e$$

exports as capacity to import as:

$$E_{cm} = E/P_m$$

imports in constant prices as:

$$M_k = E/P_m$$

---

*Based on IBRD, "The CPP System for Country Economic Projections," Chapter 6.

terms of trade income effect as:

$$TTE = E/P_m - E/P_e = E_{cm} - E_k$$

The incorporation of the terms of trade income effect into the system of national accounts becomes a simple matter. Excluding the terms of trade income effect, the resources available to a country and their use can be expressed as follows:

$$GDP + M - E = C + I \qquad (C.1)$$

To express the same relationship in constant prices, equation (C.1) can be written in the following form:

$$GDP_k + M_k - E_k = C_k + I_k \qquad (C.2)$$

To incorporate the terms of trade effect, equation (C.2) is written as follows:

$$GDY_k + M_k - E_{cm} = C_k + I_k$$

By substitution (see equation [C.2]), one obtains:

$$GDY_k + M_k - E_{cm} = GDP_k + M_k - E_k$$

$$GDY_k = GDP_k + M_k - E_k - M_k + E_{cm}$$

$$GDY_k = GDP_k + E_{cm} - E_k$$

Also by substitution, one obtains:

$$GDY_k = GDP_k + TTE$$

APPENDIX D
STATISTICAL TABLES

The following seven tables were too extensive to include in the main text.

TABLE D.1

Iraq: Current Account, 1950–74
(millions of Iraqi dinars)

	1950	1951	1952	1953	1954	1955	1956	1957	1958
Exports—oil	23.6	33.5	111.3	120.1	156.1	169.2	156.9	114.4	186.7
Exports—other	26.1	32.1	23.9	23.0	20.7	16.2	13.7	14.0	15.7
Total exports	49.7	65.6	135.2	143.1	176.8	185.4	170.6	128.4	202.4
Imports—oil	-8.4	-10.9	-16.8	-12.4	-5.8	-6.3	-7.4	-10.4	-10.0
Imports—other	-29.9	-41.1	-46.2	-56.9	-70.1	-90.9	-107.2	-112.0	-99.8
Total imports	-38.3	-52.0	-63.0	-69.3	-75.9	-97.2	-114.6	-122.4	-109.8
Balance of trade	11.4	13.6	72.2	73.8	100.9	88.2	56.0	6.0	92.6
Services—receipts	9.7	11.6	12.4	15.1	14.5	22.3	23.3	27.8	27.4
Services—payments	-22.8	-23.6	-92.1	-67.9	-83.2	-98.0	-89.7	-72.7	-100.6
Services—net	-13.1	-12.0	-79.7	-52.8	-68.7	-75.7	-66.4	-44.9	-73.2
Transfer payments—net	0.2	n.a.	n.a.	-0.8	0.1	0.6	-0.1	-0.1	—
Balance on current account	-1.5	1.6	-7.5	20.2	32.3	13.1	-10.5	-39.0	19.4

	1959	1960	1961	1962	1963	1964	1965	1966
Exports—oil	202.1	222.9	224.5	225.4	259.9	282.0	293.7	308.9
Exports—other	14.3	10.7	11.8	21.7	18.9	17.9	21.2	26.5
Total exports	216.4	233.6	236.3	247.1	278.8	299.9	314.9	335.4
Imports—oil	-17.0	-14.6	-12.1	-1.8	-1.6	-0.7	-0.9	-0.9
Imports—other	-99.4	-128.3	-138.9	-132.5	-112.2	-145.2	-162.6	-175.3
Total imports	-116.4	-142.9	-151.0	-134.3	-113.8	-145.9	-163.5	-176.2
Balance of trade	100.0	90.7	85.3	112.8	165.0	154.0	151.4	159.2
Services—receipts	19.0	20.8	20.3	17.2	17.5	28.4	32.9	34.8
Services—payments	-107.4	-117.6	-118.3	-119.4	-142.7	-168.8	-176.9	-187.5
Services—net	-88.4	-96.8	-98.0	-102.2	-125.2	-140.4	-144.0	-152.7
Transfer payments—net	-2.9	-3.4	-2.1	-1.8	0.1	1.2	0.3	-0.5
Balance on current account	8.7	-9.5	-14.8	8.8	39.9	14.8	7.7	6.0

(continued)

Table D.1 continued

	1967	1968	1969	1970	1971	1972	1973	1974
Exports—oil	272.0	345.0	347.4	378.1	526.2	421.5	625.5	1,943.0
Exports—other	23.8	25.9	25.9	24.7	23.2	30.9	32.8	34.4
Total exports	295.8	370.9	373.3	402.8	549.4	452.4	658.3	1,977.4
Imports—oil	-0.8	-1.1	-1.3	-2.2	-2.6	-8.6	-22.3	-70.4
Imports—other	-150.6	-144.0	-156.3	-180.7	-246.6	-234.5	-258.6	-835.3
Total imports	-151.4	-145.1	-157.6	-182.9	-249.2	-243.1	-280.9	-905.7
Balance of trade	144.4	225.8	215.7	219.9	300.2	209.3	377.4	1,071.7
Services—receipts	33.9	36.6	46.0	42.9	55.9	63.1	80.7	177.3
Services—payments	-158.0	-208.6	-209.4	-227.8	-288.7	-197.4	-202.2	-488.9
Services—net	-124.1	-172.0	-163.4	-184.9	-232.8	-134.3	-121.5	-311.6
Transfer payments—net	5.0	1.6	1.9	0.7	0.6	0.9	-2.6	-69.3
Balance on current account	25.3	55.4	54.2	35.7	68.0	75.9	253.3	690.8

n.a.: There were no transfer payments in those years.
Sources: Iraq, CBI, Bulletin, various issues.

TABLE D.2

Iraq: Gross Domestic Product by Economic Sector at Current Prices, 1953-76
(millions of Iraqi dinars)

Sector	1953	1954	1955	1956	1957	1958	1959	1960
Agriculture, forestry, and fishing	71.5	84.7	65.3	89.2	111.6	92.8	82.0	97.8
Oil extraction	128.9	149.5	161.2	152.4	113.1	175.4	190.0	208.1
Other mining and quarrying	0.9	0.9	1.6	1.6	1.7	1.8	1.8	1.7
Manufacturing	19.7	21.9	26.9	32.1	35.2	36.8	44.8	54.4
Construction	11.3	17.2	21.3	24.8	27.7	29.8	28.7	23.1
Electricity, water, and gas	1.5	1.8	2.2	2.5	2.7	2.8	3.0	3.6
Transport, communication, and storage	21.4	22.1	24.6	27.6	29.9	30.6	34.3	39.7
Wholesale and retail trade	17.8	20.7	21.5	26.9	29.7	27.5	26.2	32.6
Banking, insurance, and real estate	3.2	3.6	4.5	6.3	6.6	7.4	8.2	8.7
Ownership of dwellings	11.6	11.9	12.2	12.5	12.8	12.5	11.6	11.9
Public administration and defense	18.3	20.8	24.3	28.1	32.1	37.6	45.6	45.7
Other services	16.8	19.2	21.3	24.8	27.0	29.6	33.3	38.1
GDP at factor cost	322.9	374.3	386.9	428.8	430.1	484.6	509.5	565.4

(continued)

Table D.2 continued

Sector	1961	1962	1963	1964	1965	1966	1967	1968
Agriculture, forestry, and fishing	117.0	140.4	109.3	148.1	153.2	163.4	187.8	196.0
Oil extraction	209.0	210.2	242.5	266.9	281.1	298.5	265.2	334.6
Other mining and quarrying	2.2	1.9	1.9	1.9	4.8	5.7	5.4	6.4
Manufacturing	59.5	65.1	64.2	63.4	69.4	74.7	83.4	94.6
Construction	23.9	19.6	20.3	18.7	30.5	34.5	32.8	36.8
Electricity, water, and gas	5.0	5.5	5.2	7.7	12.0	12.6	12.8	14.9
Transport, communication, and storage	46.0	47.0	48.8	54.9	58.2	63.2	64.0	65.8
Wholesale and retail trade	36.6	38.6	35.9	44.5	69.8	74.7	79.7	86.9
Banking, insurance, and real estate	11.1	11.4	12.8	7.7	9.5	12.4	12.3	13.1
Ownership of dwellings	12.1	12.4	12.4	14.8	30.2	39.0	32.4	37.9
Public administration and defense	51.5	59.8	67.4	81.2	89.0	94.1	98.2	104.6
Other services	41.4	46.5	49.9	51.4	59.9	65.2	67.9	75.4
GDP at factor cost	615.3	658.4	670.6	761.2	867.6	938.0	941.9	1,067.0

Sector	1969	1970	1971	1972	1973	1974	1975	1976
Agriculture, forestry, and fishing	191.0	206.9	212.5	269.4	225.9	232.1	297.3	348.7
Oil extraction	335.9	362.6	507.8	400.0	563.4	2,022.7	2,279.0	2,446.0
Other mining and quarrying	7.3	7.9	5.1	7.3	10.9	8.0	8.7	29.1
Manufacturing	103.0	116.0	118.5	140.0	157.6	176.1	238.5	324.5
Construction	38.5	40.6	43.6	45.2	57.6	69.1	91.3	355.1
Electricity, water, and gas	16.8	17.8	11.9	13.7	16.0	13.7	17.7	22.5
Transport, communication, and storage	69.1	71.2	79.7	85.9	88.5	124.1	157.6	217.7
Wholesale and retail trade	90.1	98.6	94.4	102.6	115.2	168.9	194.9	197.8
Banking, insurance, and real estate	15.5	18.6	20.6	20.0	20.5	44.2	60.2	89.0
Ownership of dwellings	44.7	51.0	54.5	56.3	58.8	74.7	80.5	75.9
Public administration and defense	117.8	124.3	131.7	136.0	154.5	261.4	372.6	432.8
Other services	80.0	86.9	94.7	112.4	118.6	152.7	172.2	43.7
GDP at factor cost	1,109.7	1,202.4	1,375.0	1,388.8	1,587.5	3,347.7	3,970.5	4,582.8

Sources: Iraq, CSO, *Annual Abstract of Statistics 1970*; Iraq, CSO, *Annual Abstract of Statistics 1974*; Iraq, CSO, *Annual Abstract of Statistics 1976*.

TABLE D.3

Iraq: Gross Domestic Product by Economic Sector at Constant 1969 Prices, 1953-76
(millions of Iraqi dinars)

Sector	1953	1954	1955	1956	1957	1958	1959	1960
Agriculture, forestry, and fishing	90.4	112.8	79.2	104.7	129.3	105.6	90.6	108.3
Oil extraction	162.9	199.2	195.3	178.8	131.1	199.8	210.0	230.3
Other mining and quarrying	1.2	1.2	1.9	1.9	2.0	2.0	2.0	1.9
Manufacturing	24.9	29.3	32.6	37.6	40.8	42.0	49.5	60.1
Construction	14.2	22.9	25.8	29.2	32.1	34.0	31.7	25.6
Electricity, water, and gas	1.4	2.3	2.6	3.0	3.1	3.2	3.3	4.0
Transport, communication, and storage	27.0	29.4	29.8	32.3	34.7	34.9	37.8	44.0
Wholesale and retail trade	22.6	27.6	26.1	31.6	34.4	31.4	29.0	36.0
Banking, insurance, and real estate	4.0	4.9	5.4	7.3	7.6	8.5	9.0	9.6
Ownership of dwellings	13.6	15.8	14.7	14.5	14.8	14.2	12.8	13.1
Public administration and defense	23.1	27.7	29.5	33.0	37.2	42.8	50.5	50.6
Services	21.3	25.7	25.8	29.2	31.3	33.7	36.8	42.2
GDP at factor cost	406.6	498.8	468.7	503.1	498.4	552.1	563.0	625.7

Sector	1961	1962	1963	1964	1965	1966	1967	1968
Agriculture, forestry, and fishing	132.8	157.2	118.3	141.4	168.7	173.3	183.2	198.8
Oil extraction	237.2	235.4	262.4	289.6	309.6	316.6	258.7	339.4
Other mining and quarrying	2.5	2.1	2.1	2.1	5.3	6.0	5.3	6.5
Manufacturing	67.5	72.9	69.5	67.5	76.4	79.2	81.4	96.0
Construction	27.1	21.9	22.0	28.3	33.6	36.6	32.0	37.3
Electricity, water, and gas	5.7	6.2	5.6	11.8	13.2	13.4	12.5	15.1
Transport, communication, and storage	52.2	52.6	52.8	48.2	64.1	67.0	62.4	66.7
Wholesale and retail trade	41.6	43.2	38.9	65.9	76.9	79.2	77.8	88.1
Banking, insurance, and real estate	12.6	15.0	13.9	8.5	10.5	13.1	12.0	13.3
Ownership of dwellings	13.7	14.0	13.4	31.0	33.2	41.4	31.6	38.4
Public administration and defense	58.5	67.0	72.9	90.6	98.0	99.8	95.8	106.1
Services	47.0	52.1	54.0	56.8	66.0	69.1	66.2	76.5
GDP at factor cost	698.4	739.6	725.8	841.7	955.5	994.7	918.9	1,082.2

(continued)

Table D.3 continued

Sector	1969	1970	1971	1972	1973	1974	1975	1976
Agriculture, forestry, and fishing	191.0	190.3	179.7	228.6	180.5	166.1	163.5	180.7
Oil extraction	335.9	312.5	377.9	326.2	448.1	438.1	509.6	541.2
Other mining and quarrying	7.3	7.4	4.8	6.8	9.7	5.4	5.8	18.1
Manufacturing	103.0	117.9	122.5	132.6	143.5	153.7	182.4	228.5
Construction	38.5	38.9	40.0	42.3	51.1	52.7	69.2	248.3
Electricity, water, and gas	16.8	12.7	11.9	13.7	16.0	18.6	23.4	35.2
Transport, communication, and storage	69.1	69.4	75.8	75.8	74.2	93.5	122.9	156.6
Wholesale and retail trade	90.1	94.8	87.3	90.1	96.5	130.7	137.2	129.3
Banking, insurance, and real estate	15.5	17.6	18.4	17.6	17.2	34.1	42.6	58.6
Ownership of dwellings	44.7	49.2	50.8	52.2	53.7	65.4	68.5	58.8
Public administration and defense	117.8	124.1	131.3	135.6	154.1	215.5	310.0	307.0
Services	80.0	85.4	91.5	109.0	115.4	125.4	139.4	32.4
GDP at factor cost	1,109.7	1,120.2	1,191.9	1,230.5	1,360.0	1,499.2	1,774.5	1,994.7

Sources: Jawad Hashim, Hussein Omar, and Ali al-Munoufy, Evaluation of Economic Growth in Iraq, 1950–1970, vol. 1, The Planning Experience (Baghdad, Iraq: Ministry of Planning, 1970), p. 286; IBRD, Economic and Social Data Bank; and Iraq, CSO, Annual Abstract of Statistics 1976.

TABLE D.4

Iraq: Gross Domestic Fixed Capital Formation by Economic Sector at Current Prices, 1953-75
(millions of Iraqi dinars)

Sector	1953	1954	1955	1956	1957	1958	1959	1960
Agriculture, forestry, and fishing	10.2	10.9	11.4	12.1	18.4	15.6	12.5	12.7
Mining and quarrying	9.2	9.8	10.3	10.8	5.0	5.9	15.1	23.1
Manufacturing	10.2	10.8	11.3	11.9	16.7	13.5	11.3	9.0
Construction	1.4	1.5	1.6	1.7	2.9	2.1	1.8	1.3
Electricity and water	4.5	4.8	5.1	5.3	6.4	8.9	5.0	7.8
Transport, communication, and storage	18.3	19.3	20.4	21.5	23.6	21.9	20.7	25.5
Wholesale and retail trade	1.9	2.0	2.1	2.2	3.2	2.1	2.2	2.2
Banking and insurance	0.6	0.6	0.6	0.7	0.8	1.0	0.8	0.5
Ownership of dwellings	15.8	16.7	17.7	18.6	19.0	17.0	22.3	23.1
Public administration	2.5	2.6	2.8	2.9	3.1	2.4	2.7	3.2
Services	7.6	8.0	8.5	8.9	7.2	7.4	9.2	11.9
Total	82.2	87.0	91.8	96.6	106.3	97.8	103.6	120.3

(continued)

Table D.4 continued

Sector	1961	1962	1963	1964	1965	1966	1967	1968
Agriculture, forestry, and fishing	14.9	11.2	8.7	9.9	11.4	15.8	15.8	16.8
Mining and quarrying	22.8	4.9	1.8	0.9	1.2	1.5	1.5	1.2
Manufacturing	12.8	20.9	21.1	22.4	25.4	30.6	34.3	36.4
Construction	1.9	1.7	1.1	1.3	1.9	2.4	2.5	1.7
Electricity and water	5.0	4.9	7.3	11.6	10.6	11.5	12.5	8.7
Transport, communication, and storage	32.1	28.9	27.1	26.6	24.5	31.0	24.7	20.6
Wholesale and retail trade	3.0	2.9	3.0	3.8	4.3	4.7	4.7	7.4
Banking and insurance	0.9	1.3	0.7	0.8	0.8	0.8	1.0	1.1
Ownership of dwellings	25.8	24.2	21.7	24.9	27.6	31.9	24.1	26.0
Public administration	4.1	5.0	3.8	3.8	3.7	3.7	5.9	4.6
Services	13.9	13.3	11.1	16.1	18.4	15.7	16.7	18.5
Total	137.2	119.2	107.4	122.1	129.8	149.6	143.7	143.0

Sector	1969	1970	1971	1972	1973	1974[a]	1975[b]
Agriculture, forestry, and fishing	22.5	23.0	29.0	31.3	33.9	47.8	50.4
Mining and quarrying	1.1	7.6	10.4	13.2	30.7	79.6	117.8
Manufacturing	40.1	42.5	43.4	50.3	69.1	123.7	216.2
Construction	4.0	3.9	5.1	5.0	7.8	21.3	31.9
Electricity and water	8.5	12.1	11.0	10.7	9.8	7.3	13.2
Transport, communication, and storage	18.2	26.8	27.7	31.0	32.5	84.8	98.7
Wholesale and retail trade	5.4	7.9	6.3	7.0	22.6	25.0	42.5
Banking and insurance	1.3	1.5	2.5	1.3	6.1	1.3	2.4
Ownership of dwellings	32.3	32.8	34.4	39.2	45.1	45.3	67.5
Public administration	4.0	5.2	5.4	5.9	12.3	41.1	57.5
Services	19.8	21.8	19.5	22.1	18.7	54.7	63.1
Total	157.2	185.1	194.7	217.0	288.6	531.9	761.2

[a]Revised.
[b]Provisional.

Sources: Jawad Hashim, Hussein Omar, and Ali al-Munoufy, Evaluation of Economic Growth in Iraq, 1950–1970, vol. 1, The Planning Experience (Baghdad, Iraq: Ministry of Planning, 1970); Jawad Hashim, Fixed Capital Formation in Iraq: 1957–1970 (Baghdad, Iraq: Ministry of Planning, 1972); Iraq, CSO, Annual Abstract of Statistics 1974 (1957–1970); and Iraq, CSO, Annual Abstract of Statistics 1976.

TABLE D.5

Iraq: Gross Domestic Fixed Capital Formation by Economic Sector at Constant 1969 Prices, 1953-75
(millions of Iraqi dinars)

Sector	1953	1954	1955	1956	1957	1958	1959	1960
Agriculture, forestry, and fishing	12.6	14.2	13.5	13.9	22.1	19.0	14.0	13.8
Mining and quarrying	11.4	12.8	12.2	12.4	5.0	7.2	17.0	25.2
Manufacturing	12.6	14.1	13.4	13.7	20.4	16.5	12.9	10.4
Construction	1.7	1.9	1.9	1.9	3.6	2.6	2.1	1.6
Electricity and water	5.5	6.3	6.0	6.0	7.7	10.9	5.7	8.6
Transport, communication, and storage	22.6	25.2	24.2	24.6	28.8	27.2	23.3	28.3
Wholesale and retail trade	2.3	2.6	2.4	2.5	3.8	2.5	2.5	2.4
Banking and insurance	0.8	0.8	0.8	0.8	0.9	0.8	0.9	0.6
Ownership of dwellings	19.5	21.2	21.0	21.3	22.4	20.7	25.0	24.7
Public administration	3.1	3.9	3.3	3.3	3.7	3.0	3.0	3.5
Services	9.3	10.4	10.1	10.2	8.5	8.9	10.3	13.0
Total	101.4	113.4	108.8	110.6	126.9	119.3	116.7	132.1

Sector	1961	1962	1963	1964	1965	1966	1967	1968
Agriculture, forestry, and fishing	16.7	12.5	9.7	10.8	12.5	16.9	16.6	17.0
Mining and quarrying	25.3	5.6	1.9	1.0	1.2	1.6	1.6	1.2
Manufacturing	14.6	23.4	23.5	24.4	27.7	32.7	35.4	36.8
Construction	2.1	1.9	1.2	1.5	2.0	2.5	2.5	1.7
Electricity and water	5.6	5.5	8.2	12.5	11.7	12.3	13.1	8.9
Transport, communication, and storage	35.8	32.4	30.6	28.0	25.5	29.9	23.0	20.2
Wholesale and retail trade	3.4	3.2	3.4	4.0	4.8	4.9	5.6	7.6
Banking and insurance	1.0	1.3	0.8	0.8	0.8	0.9	1.0	1.1
Ownership of dwellings	28.8	27.1	24.6	27.1	30.4	34.4	25.5	26.8
Public administration	4.5	5.7	4.3	4.1	4.0	3.8	6.0	4.7
Services	15.4	14.8	12.5	17.7	20.5	17.0	17.6	19.1
Total	152.7	133.4	120.7	131.9	141.1	156.9	147.9	145.1

(continued)

Table D.5 continued

Sector	1969	1970	1971	1972	1973	1974[a]	1975[b]
Agriculture, forestry, and fishing	22.5	22.5	28.1	29.8	31.6	43.8	45.5
Mining and quarrying	1.1	7.4	10.1	12.5	28.6	72.9	106.1
Manufacturing	40.1	41.4	38.5	47.6	64.6	113.8	195.8
Construction	4.0	3.8	4.8	4.7	7.3	19.7	28.9
Electricity and water	8.5	11.9	10.6	10.2	9.1	6.6	11.9
Transport, communication, and storage	18.2	25.5	26.3	28.9	30.4	78.6	90.3
Wholesale and retail trade	5.4	7.6	6.0	6.6	21.2	23.2	38.8
Banking and insurance	1.3	1.5	2.5	1.2	5.7	1.2	2.2
Ownership of dwellings	32.3	32.3	33.5	37.5	41.8	41.4	60.8
Public administration	4.0	5.0	5.2	5.5	11.5	37.9	52.2
Services	19.8	21.4	18.9	21.2	17.4	50.1	57.0
Total	157.2	180.3	184.5	205.7	269.2	489.2	689.5

[a]Revised.
[b]Provisional.

Sources: Jawad Hashim, Hussein Omar, and Ali al-Munoufy, Evaluation of Economic Growth in Iraq, 1950-1970, vol. 1, The Planning Experience (Baghdad, Iraq: Ministry of Planning, 1970); Jawad Hashim, Fixed Capital Formation in Iraq: 1957-1970 (Baghdad, Iraq: Ministry of Planning, 1972); Iraq, CSO, Annual Abstract of Statistics 1974 (1957-1957-1970); and Iraq, CSO, Annual Abstract of Statistics 1976.

TABLE D.6

Results of Ordinary Least Squares Estimations of Aggregate
and Sectoral Growth Rates of the Iraqi Economy

	ln A	ln B	R^2	SE[a]	Rate of Growth
Overall GDP					
1954–73	6.054 (230.450)[b]	0.059 (26.696)	0.975	0.057	6.029
1964–73	6.722 (223.244)	0.046 (9.507)	0.919	0.044	4.722
Nonoil GDP					
1954–73	5.571 (183.085)	0.065 (25.757)	0.974	0.065	6.761
1964–73	6.320 (244.900)	0.051 (12.218)	0.949	0.038	5.213
Nonoil, nonagricultural GDP					
1954–73	5.164 (45.760)	0.068 (7.254)	0.745	0.243	7.073
1964–73	5.994 (234.134)	0.058 (14.051)	0.961	0.037	5.968
GDP in the agricultural sector					
1954–73	4.496 (68.913)	0.044 (8.156)	0.787	0.140	4.542
1964–73	5.053 (77.625)	0.028 (2.646)	0.467	0.095	2.814
GDP in the mining and quarrying sector[c]					
1954–73	5.091 (85.728)	0.046 (9.227)	0.825	0.128	4.680
1964–73	5.566 (83.061)	0.037 (3.775)	0.613	0.104	3.801
GDP in the industrial sectors[d]					
1954–73	4.017 (179.660)	0.065 (34.668)	0.985	0.048	6.685
1964–73	4.631 (184.773)	0.070 (17.226)	0.974	0.037	7.206
GDP in the distribution sectors[e]					
1954–73	4.057 (100.105)	0.066 (19.625)	0.955	0.087	6.865
1964–73	4.893 (118.763)	0.039 (5.849)	0.810	0.060	3.960
GDP in the services sectors[f]					
1954–73	4.112 (141.277)	0.086 (35.302)	0.986	0.063	8.956
1964–73	5.114 (180.234)	0.063 (13.845)	0.960	0.042	6.536

[a]Standard error of estimation.
[b]Values in parentheses are T-ratios.
[c]Oil extraction and other mining and quarrying.
[d]Manufacturing; construction; and electricity, water, and gas.
[e]Transport, communications, and storage; wholesale and retail trade; and banking, insurance, and real estate.
[f]Ownership of dwellings, public administration and defense, and other services.
Source: Calculated from data in Table D.3.

TABLE D.7

Results of Ordinary Least Squares Estimations of Aggregate and Sectoral Growth Rates of Gross Domestic Fixed Capital Formation in Iraq

	ln A	ln B	R^2	SE[a]	Rate of Growth (percent)
Overall GDFC					
1953-73	4.576	0.034	0.80	0.11	3.47
	(93.856)[b]	(8.789)			
1964-73	4.717	0.063	0.85	0.09	6.50
	(79.281)	(7.183)			
1954-58	4.611	0.034	0.63	0.05	3.46
	(91.754)	(2.636)			
1959-63	4.796	0.017	0.10	0.11	1.74[c]
	(48.342)	(0.678)			
1964-68	4.805	0.039	0.62	0.06	4.00
	(81.426)	(2.583)			
1969-73	4.836	0.112	0.92	0.07	11.85
	(76.855)	(6.931)			
GDFC in public sector					
1953-73	3.822	0.044	0.73	0.17	4.46
	(49.931)	(7.151)			
1964-73	4.039	0.077	0.67	0.19	8.00
	(33.364)	(4.312)			
1954-58	3.837	0.063	0.74	0.08	6.50
	(52.605)	(3.364)			
1959-63	4.118	0.006	0.03	0.08	0.60[c]
	(58.389)	(0.333)			
1964-68	4.186	0.040	0.63	0.06	4.12
	(69.396)	(2.606)			
1969-73	4.066	0.167	0.77	0.19	18.15
	(23.040)	(3.680)			
GDFC in private sector					
1953-73	3.968	0.021	0.49	0.13	2.08
	(65.425)	(4.270)			
1964-73	4.056	0.036	0.48	0.13	3.62
	(49.293)	(2.927)			
1954-58	4.004	0.002	0.01	0.05	0.21[c]
	(78.997)	(0.162)			
1959-63	4.072	0.030	0.08	0.21	3.07[c]
	(21.071)	(0.610)			
1964-68	4.032	0.038	0.46	0.08	3.82
	(50.431)	(1.827)			
1969-73	4.318	0.013	0.03	0.16	1.33[c]
	(28.181)	(0.336)			

	ln A	ln B	R^2	SE^a	Rate of Growth (percent)
GDFC in agriculture					
1954-73	2.455	0.033	0.38	0.27	3.33
	(20.240)	(3.390)			
1964-73	2.190	0.121	0.97	0.08	12.82
	(44.901)	(16.774)			
1954-58	2.403	0.097	0.67	0.14	10.23
	(17.951)	(2.835)			
1959-63	2.988	-0.100	0.65	0.15	-9.54
	(20.725)	(-2.710)			
1964-68	2.168	0.126	0.89	0.09	13.38
	(24.895)	(5.617)			
1969-73	2.791	0.119	0.92	0.07	12.64
	(40.684)	(6.756)			
GDFC in industry					
1954-73	2.836	0.066	0.86	0.16	6.78
	(38.405)	(11.150)			
1964-73	3.496	0.069	0.89	0.08	7.20
	(64.333	(8.673)			
1954-58	2.865	0.090	0.73	0.12	9.41
	(26.614)	(3.252)			
1959-63	3.074	0.050	0.18	0.22	5.08[c]
	(15.122)	(0.948)			
1964-68	3.471	0.080	0.84	0.07	8.37
	(51.674)	(4.662)			
1969-73	3.750	0.090	0.81	0.09	9.38
	(44.873)	(4.179)			

[a] Standard error of estimation.

[b] Values in brackets are T-ratios.

[c] Not significant.

Source: Calculated from data in Table 5.2 and Table D.5.

BIBLIOGRAPHY

BOOKS

Adelman, Irma, and Erik Thorebecke, eds. The Theory and Design of Economic Development. Baltimore: Johns Hopkins University Press, 1966.

Adelman, M. A. The World Petroleum Market. Baltimore: Johns Hopkins University Press, 1972.

Adler, John H. Absorptive Capacity, the Concept and Its Determinants. Washington, D.C.: Brookings Institution, 1965.

Alnasrawi, Abbas. Financing Economic Development in Iraq. New York: Praeger, 1967.

Aschheim, Joseph, and Ching-Yao Hsieh. Macroeconomics: Income and Monetary Theory. Columbus, Ohio: Charles E. Merrill, 1969.

Berrill, Kenneth, ed. Economic Development with Special Reference to East Asia. London: Macmillan, 1965.

Bilas, Richard A. Microeconomic Theory: A Graphical Analysis. New York: McGraw-Hill, 1967.

Cooper, Charles A., and Sidney S. Alexander, eds. Economic Development and Population Growth in the Middle East. New York: American Elsevier, 1972.

Ghonemy, Mohamad Riad el-, ed. Land Policy in the Near East. Rome: Food and Agriculture Organization of the United Nations, 1967.

Hashim, Jawad. Fixed Capital Formation in Iraq 1957-1970. Baghdad, Iraq: Ministry of Planning, 1972.

Hashim, Jawad, Hussein Omar, and Ali al-Munoufy. Evaluation of Economic Growth in Iraq, 1950-1970. Vol. 1. The Planning Experience. Vol. 2. The Evolution of the Commodities Sectors. Baghdad, Iraq: Ministry of Planning, 1970.

Hirschman, Albert O. The Strategy of Economic Development. New Haven, Conn.: Yale University Press, 1958.

Houthakker, Hendrik S. The World Price of Oil: A Medium Term Analysis. Washington, D.C.: American Enterprise Institute for Public Policy Research, 1976.

Hughes, Helen, Donald B. Keeging, Karsten Laursen, Goran Ohlin, and John D. Schilling. Prospects for Developing Countries, 1978-85. Washington, D.C.: IBRD, 1977.

IBRD. The Economic Development of Iraq. Baltimore: Johns Hopkins University Press, 1952.

Jalal, Ferhang. The Role of Government in the Industrialization of Iraq. London: Frank Cass, 1972.

Kanaan, Taher H. Input-Output and Social Accounts of Iraq 1960-1963. Baghdad, Iraq: Ministry of Planning, 1965.

Langley, Kathleen M. The Industrialization of Iraq. Cambridge, Mass.: Harvard University Press, 1962.

Levine, Jonathan V. The Export Economies: Their Pattern of Development in Historical Perspective. Cambridge, Mass.: Harvard University Press, 1960.

Little, Ian, Tibor Scitovsky, and Maurice Scott. Industry and Trade in Some Developing Countries. New York: Oxford University Press, 1970.

Little, Inc., Arthur D. A Plan for Industrial Development in Iraq. Cambridge, Mass.: Arthur D. Little, 1956.

Longrigg, Stephen H. Oil in the Middle East. 3d ed. London: Oxford University Press, 1968.

Meier, Gerald E. Leading Issues in Economic Development. New York: Oxford University Press, 1970.

Mikdashi, Zuhayr. A Financial Analysis of Middle Eastern Oil Concessions: 1901-1965. New York: Praeger, 1966.

Mikdashi, Zuhayr, Sherill Cleland, and Ian Seymour, eds. Continuity and Change in the World Oil Industry. Beirut, Lebanon: Middle East Research and Publishing Center, 1970.

Mosawi, Muhsen al-. Iraq's Oil: The People's Struggle against Oil Companies Covets. Baghdad, Iraq: Ministry of Information, 1973.

Myint, Hla. The Economics of the Developing Countries. New York: Praeger, 1965.

Nurkse, Ragnar. Problems of Capital Formation in Underdeveloped Countries. Oxford: Basil Blackwell, 1953.

Ølgaard, Anders. Growth, Productivity and Relative Prices. Amsterdam: North-Holland, 1966.

Phelps, Edmund S., ed. The Goal of Economic Growth. New York: W. W. Norton, 1962.

Rostow, W. W., ed. The Stages of Economic Growth. New York: Cambridge University Press, 1960.

―――. Economics of Take-Off into Sustained Growth. New York: St. Martin's Press, 1963.

Salter, Lord (James Arthur). The Development of Iraq: A Plan of Action. Baghdad: Iraq Development Board, 1955.

Shwadran, Benjamin. The Middle East, Oil and the Great Powers. 2d ed. rev. New York: Council for Middle Eastern Affairs Press, 1959.

Sodersten, Bo. International Economics. New York: Harper and Row, 1970.

Stolper, Wolfgang F. Planning without Facts, Lessons in Resource Allocation from Nigeria's Development. Cambridge, Mass.: Harvard University Press, 1966.

Theberge, James D., ed. Economics of Trade and Development. New York: John Wiley and Sons, 1968.

Vanek, Jaroslav. Estimating Foreign Resource Needs for Economic Development. New York: McGraw-Hill, 1967.

Warriner, Doreen. Land Reform and Development in the Middle East. 2d ed. London: Oxford University Press, 1962.

―――. Land Reform in Principle and Practice. Oxford: Clarendon Press, 1969.

Waterston, Albert. Development Planning: Lessons of Experience. Baltimore: Johns Hopkins University Press, 1965.

ARTICLES

Abu el-Haj, Ribhi. "Capital Formation in Iraq, 1922-1957." Economic Development and Cultural Change 9 (1961): 604-17.

Badre, Albert Y. "Economic Development in Iraq." In Economic Development and Population Growth in the Middle East, edited by Charles A. Cooper and Sidney A. Alexander, pp. 283-328. New York: American Elsevier, 1972.

Baran, P. A., and E. J. Hobsbawm. "The Stages of Economic Growth." Kyklos 14 (1961): 234-43.

Bauer, P. T., and Charles Wilson. "The Stages of Growth." Economica 29 (1962): 190-200.

Berrill, Kenneth. "Historical Experience: The Problem of Economic 'Take-Off.'" In Economic Development with Special Reference to East Asia, edited by Kenneth Berrill, pp. 233-45. London: Macmillan, 1965.

Broderick, J. B. "National Accounts at Constant Prices." The Review of Income and Wealth 13 (1967): 247-58.

Celasun, Merith, and Frank Pinto. "Energy Prospects in OECD Countries and Possible Demand for OPEC Oil Exports to 1980." World Bank Staff Working Paper, No. 221, September 1975. Mimeographed.

Checkland, S. G. "Theories of Economic and Social Evolution: The Rostow Challenge." Scottish Journal of Political Economy 7 (1960): 169-93.

Chenery, H. B. "The Application of Investment Criteria." Quarterly Journal of Economics 67 (1953): 76-96.

Chenery, H. B., and A. MacEwan. "Optimal Patterns of Growth and Aid: The Case of Pakistan." In The Theory and Design of Economic Development, edited by Irma Adelman and Erik Thorbecke, pp. 149-78. Baltimore: Johns Hopkins University Press, 1966.

Chenery, Hollis B., and Alan M. Strout. "Foreign Assistance and Economic Development." The American Economic Review 56 (1966): 679-733.

Courbis, Raymond. "Comment on Y. Kurabayashi: The Impact of Changes in the Terms of Trade on a System of National Accounts." The Review of Income and Wealth 18 (1972): 247-50.

Currie, Lauchlin. "The 'Leading Sector' Model of Growth in Developing Countries." Journal of Economic Studies 1 (1974): 1-16.

de Vries, Rimmer. "The Build-Up of OPEC Funds." World Financial Markets, September 1974, pp. 1-10.

Fishlow, Albert. "Empty Economic Stages?" Economic Journal 75 (1965): 112-25.

Hadithy, Abdul Jalil el-, and Ahmad el-Dujaili. "Problems of Implementation of Agrarian Reform in Iraq." In Land Policy in the Near East, edited by Mohamad Riad el-Ghonemy, pp. 218-34. Rome, Italy: FAO, 1967.

Hanson, James A. "The Leading Sector Development Strategy and the Importance of Institutional Reform: A Reinterpretation." Journal of Economic Studies 3 (1976): 1-12.

Horvat, Branko. "The Optimum Rate of Investment." The Economic Journal 68 (1958): 747-67.

Iskander, Marwan. "Economic Development Plans in Oil Exporting Countries and Their Implications for Oil Production Targets." In Continuity and Change in the World Oil Industry, edited by Zuhayr M. Mikdashi, Sherrill Cleland, and Ian Seymour, pp. 39-63. Beirut, Lebanon: Middle East Research and Publishing Center, 1970.

Kahn, A. E. "Investment Criteria in Development Programs." Quarterly Journal of Economics 65 (1951): 38-61.

Kennedy, M. "An Economic Model of the World Oil Market." Bell Journal of Economics and Management Science 5 (1974): 540-77.

Kurabayashi, Yoshimasa. "The Impact of Changes in Terms of Trade on a System of National Accounts: An Attempted Synthesis." The Review of Income and Wealth 17 (1971): 285-97.

Lange, Oscar. "Economic Development, Planning and International Cooperation." In *Leading Issues in Economic Development*, edited by Gerald E. Meier, pp. 695-700. 2d ed. New York: Oxford University Press, 1970.

Lewis, W. Arthur. "On Assessing a Development Plan." *Economic Bulletin of the Economic Society of Ghana*, (May-June 1959). In *Leading Issues in Economic Development*, edited by Gerald E. Meier, pp. 716-22. 2d ed. New York: Oxford University Press, 1970.

Morgan, David R. "Fiscal Policy in Oil Exporting Countries, 1972-78." *IMF Staff Papers* 26 (1979): 55-86.

Myrdal, Gunnar. "Growth and Social Justice." *World Development* 1 (1973): 119-20.

Ndongko, Wilfred A. "'Balanced'" versus 'Unbalanced' Growth as International Development Strategies." *Mondes en Développement* 13 (1976): 904-10.

North, D. C. "A Note on Professor Rostow's 'Take-Off' into Self-Sustained Economic Growth." *The Manchester School of Economic and Social Studies* 26 (1958): 68-75.

Ohlin, Goran. "Reflections on the Rostow Doctrine." *Economic Development and Cultural Change* 11 (1961): 648-55.

Rosovsky, Henry. "The Take-Off into Sustained Controversy." *The Journal of Economic History* 25 (1965): 271-75.

Rostow, W. W. "Leading Sectors and Take-Off." In *Economics of Take-Off into Sustained Growth*, edited by W. W. Rostow, pp. 1-21. New York: St. Martin's Press, 1963.

Schatz, Sayre P. "The Capital Shortage Illusion: Government Lending in Nigeria." *Oxford Economic Papers* 17 (1965): 309-16.

Schultz, Theodore W. "Investment in Human Capital." In *The Goal of Economic Growth*, edited by Edmund S. Phelps, pp. 106-20. New York: W. W. Norton, 1962.

Sibahi, Aziz. "The Relationship between the Size of Agricultural Land and Increased Production." *Ath-Thawra*, October 20, 1972.

Singer, Hans. "The Distribution of Gains between Investing and Borrowing Countries." American Economic Review 40 (1950): 473-85.

Streeten, Paul. "Balanced versus Unbalanced Growth." The Economic Weekly, April 20, 1963, pp. 669-71.

Stuvel, G. "Asset Revaluation and Terms-of-Trade Effects in the Framework of the National Accounts." The Economic Journal 64 (1959): 275-87.

United Nations. Department of Economics and Social Affairs. "Use of Models in Programming." Industrialization and Productivity Bulletin, April 4, 1961. In Leading Issues in Economic Development, edited by Gerald E. Meier, pp. 688-95. 2d ed. New York: Oxford University Press, 1970.

United Nations. ECAFE. "Criteria for Allocating Investment Resources among Various Fields of Development in Underdeveloped Countries." Economic Bulletin for Asia and the Far East, June 1961. In Leading Issues in Economic Development, edited by Gerald E. Meier, pp. 340-44. 2d ed. New York: Oxford University Press, 1970.

Vernon, Raymond. "Foreign-Owned Enterprise in the Developing Countries." In Economics of Trade and Development, edited by James D. Theberge, pp. 446-63. New York: John Wiley and Sons, 1968.

OFFICIAL AND INTERNATIONAL
DOCUMENTS AND REPORTS

Arab League. Hashim, Jawad M. "The Structure of Fixed Capital Formation by Foreign Oil Companies Operating in Iraq during 1957-1969 and Its Contribution to the Iraqi Economy." Eighth Arab Petroleum Congress, Algiers, May 28-June 3, 1972.

IBRD. "The CPP System for Country Economic Projections." Washington, D.C.: World Bank, November 1973. Mimeographed.

_____. Fourth Annual Report to the Board of Governors: 1948-1949.

IMF. Annual Report 1977.

_____. International Financial Statistics. Various issues.

_____. IMF Survey. Various issues.

Iraq. Central Bank of Iraq. Bulletin. Various issues.

Iraq. Central Statistical Organization. Annual Abstract of Statistics. Various issues.

Iraq. Iraqi National Oil Company. Annual Review 1972.

Iraq. Ministry of Industry. Annual Report for Fiscal Year 1973-74.

Iraq. Ministry of Oil and Minerals. The Nationalization of Iraq Petroleum Company's Operations in Iraq: The Facts and the Causes. Baghdad, Iraq, 1973.

Iraq. Ministry of Planning. The Investment Program for the Year 1974-75. Baghdad, Iraq, 1975.

_____. The Investment Program for the Year 1975.

_____. Results of the Follow-Up of the Implementation of Investment Objectives of Economic Programs and Plans in Iraq: 1951-1971. 1972.

Iraq, Republic of. Public Laws. Various laws for the years 1952-76.

OPEC. Annual Statistical Bulletin 1974. Vienna, 1975.

United Nations. Department of Economic and Social Affairs. Statistical Office. World Energy Supplies: 1950-1974 (ST/ESA/STAT/SER. J/19).

_____. World Energy Supplies: 1961-1970 (ST/STAT/SER. J/15).

_____. World Energy Supplies: 1971-1975 (ST/ESA/STAT/SER. J/20).

United Nations. Economic and Social Office in Beirut. Studies on Selected Development Problems in Various Countries in the Middle East. New York: United Nations, 1967.

188 / OIL REVENUES & ACCELERATED GROWTH

PERIODICALS AND NEWSPAPERS

Al-Jumhuriyah. Various issues, 1975/1976.

Al-Naft Wal-Aalam. Various issues, 1973/1974.

Ath-Thawra. Various issues, 1974/1975.

Business Week. August 4, 1975.

The Middle East. December 1977.

Middle East Economic Survey. Various issues, 1970-79.

Mideast Markets. March 14 and 28, 1977.

Oil and Gas Journal. December 1, 1975; and February 26, 1979.

Petroleum Intelligence Weekly. June 10, 1975.

Washington Post. August 7, 1978; and May 6, 1979.

ABOUT THE AUTHOR

KADHIM A. AL-EYD is alternate executive director of the International Monetary Fund. His experience includes serving as an attaché at the Embassy of the Republic of Iraq in Washington, D.C., counselor at the Permanent Mission of Iraq to the United Nations in New York, and assistant to executive director in the World Bank and the International Monetary Fund.

Dr. Al-Eyd holds a B.S. degree from the University of Wisconsin at Madison and an M.B.A. and Ph.D. from George Washington University.